Sarah Cooper

Animal life in the sea and on the land

Sarah Cooper

Animal life in the sea and on the land

ISBN/EAN: 9783337815110

Printed in Europe, USA, Canada, Australia, Japan

Cover: Foto ©ninafisch / pixelio.de

More available books at **www.hansebooks.com**

ANIMAL LIFE

IN THE SEA AND ON THE LAND

A ZOOLOGY FOR YOUNG PEOPLE

By SARAH COOPER

ILLUSTRATED

NEW YORK

HARPER & BROTHERS, FRANKLIN SQUARE

1887

TO

THE ENTHUSIASTIC CLASSES

FOR WHICH THE LESSONS WERE PREPARED

THIS BOOK

Is Affectionately Dedicated

PREFACE.

THIS book is offered to young people with the hope that it may help them in their studies of natural history. The pleasure of every ramble in the country or by the seaside is increased by an acquaintance with the animals and plants which are found by the way, and consequently these studies bring their own reward.

It is far more charming to gain this knowledge from the objects themselves than from merely reading about them in books; and it is therefore hoped that each subject which is treated in these pages will be studied from specimens actually in hand, whenever it is possible to obtain them.

The habit of collecting natural objects and curiosities is a helpful one; and if young students are careful to find out all they can about these objects, the collection will in time represent an unexpected amount of positive knowledge.

The aim has been to make this little book accurate, and to bring it up to the present condition of science; at the same time scientific terms have been avoided when others could be substituted for them. Classification has not been made prominent, yet the arrangement of Nicholson has been adhered to throughout.

Starting with the sponge, and going systematically through the animal kingdom, a gradual development has been traced from the simple forms of life up to the highest, and such subjects have been selected for the purpose as are probably of most general interest.

Especial attention has been given to the structure of animals, and to the wonderful adaptation of this structure to their various habits and modes of life.

S. C.

PHILADELPHIA, *June*, 1887.

CONTENTS.

ILLUSTRATIONS.

ANIMAL LIFE

IN

THE SEA AND ON THE LAND.

I.

SPONGES.

SUB-KINGDOM, PROTOZOA : CLASS, RHIZOPODA.

1. Sponges.—Sponges are so common and so familiar that many of us have used them all our lives without stopping to admire their curious and interesting structure, or to inquire into the history of their past lives. We may, indeed, have noticed that they can be squeezed into a very small space, and that they will return to their natural shape when the pressure is removed. Perhaps we have also noticed that they are full of little holes or pores, and that they will absorb a large quantity of water.

Fig. 1.—SPONGE.

2. Are Sponges Animals or Vegetables?—You know there has been a doubt whether sponges belong to the animal or to the vegetable kingdom. For a long time naturalists

1

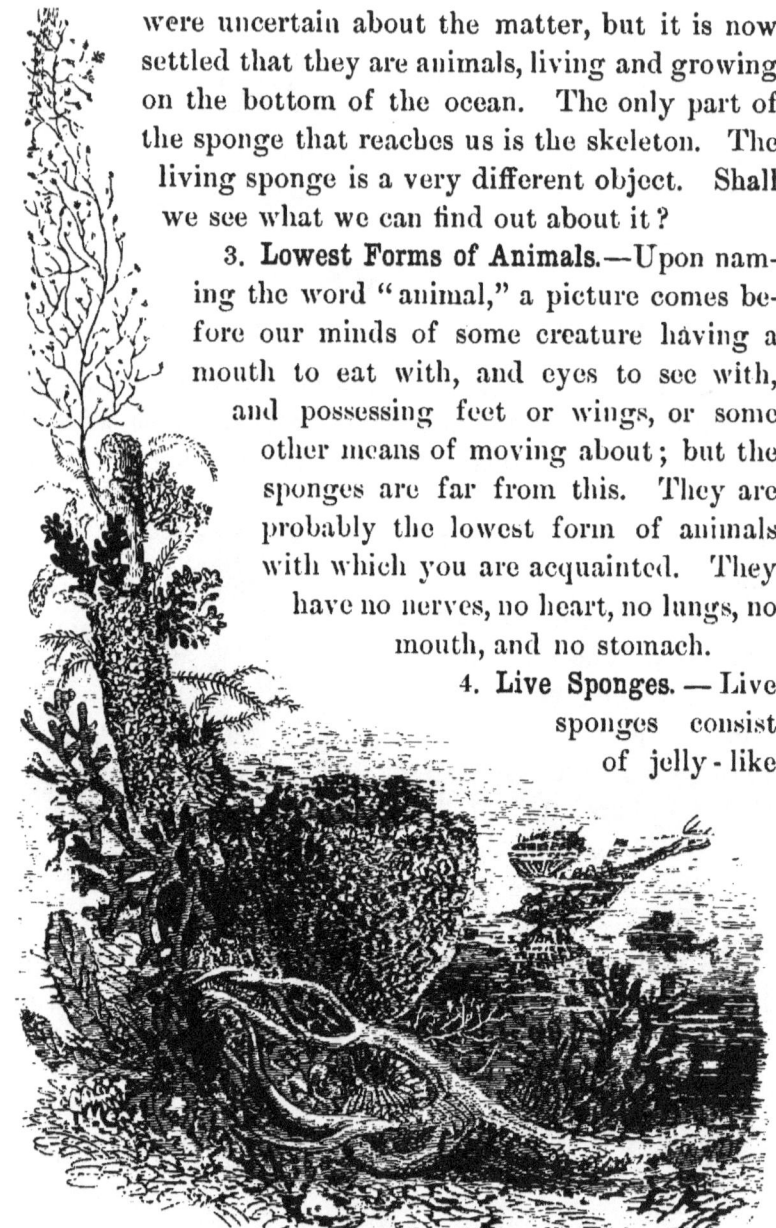

were uncertain about the matter, but it is now settled that they are animals, living and growing on the bottom of the ocean. The only part of the sponge that reaches us is the skeleton. The living sponge is a very different object. Shall we see what we can find out about it?

3. **Lowest Forms of Animals.**—Upon naming the word "animal," a picture comes before our minds of some creature having a mouth to eat with, and eyes to see with, and possessing feet or wings, or some other means of moving about; but the sponges are far from this. They are probably the lowest form of animals with which you are acquainted. They have no nerves, no heart, no lungs, no mouth, and no stomach.

4. **Live Sponges.** — Live sponges consist of jelly - like

Fig. 2.—Sponges Growing.

bodies united in a mass, and supported by a framework of horny fibres, and needle-shaped objects called "spicules," which you will see in Fig. 3, and which we must examine further after a while. This jelly-like flesh, covering all parts of the skeleton, is about as thick as the white of an egg, and it decays immediately after the death of the sponge. During life the flesh presents many bright colors; in some species it is of a brilliant green, while in others it is orange, red, or yellow.

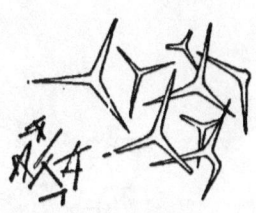

Fig. 3.—GROUPS OF SPICULES.

5. **Framework.** — The framework varies in different kinds of sponges. In those which are valuable for our use this framework consists of horny fibres interwoven in all directions until they form a mass of lacy net-work. This you can easily see with the naked eye, but by looking through a microscope you will see beauty you had not imagined. In our ordinary sponges these fibres are all that remain of the former living animal, the soft flesh with which they were covered having been removed. It is found that the horny fibres are composed of a substance very similar to the silk of a silk-worm's cocoon. They are exceedingly tough and durable.

6. **Use of Pores.** — In looking at any sponge you will notice large holes through it, with many small pores scattered between them. The living sponge draws in water through these small pores, and countless streams are continually flowing through every part of the sponge, bringing in little particles of food, and all the air it needs for breathing purposes.

7. **Cilia, and the Currents they produce.**—In order that we may understand the curious circulation in sponges, let

us examine Fig. 4, which shows a small section of a sponge with its branching canals.   One large hole is shown at *d* and the smaller pores at *b*, while in those cup-shaped hollow

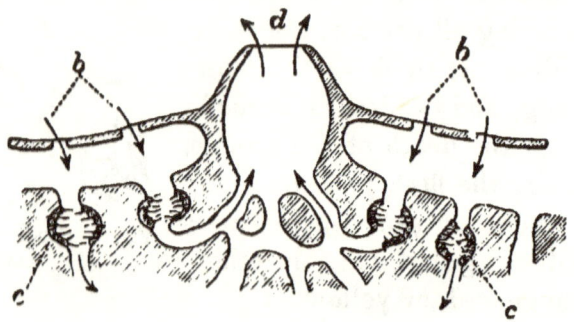

Fig. 4.—CIRCULATION OF WATER THROUGH THE SPONGE.

places in the canals marked *c* we may see a number of fine threads, or "cilia."   The word cilia means "eyelashes," but we must not mistake these threads for hairs like our eye-

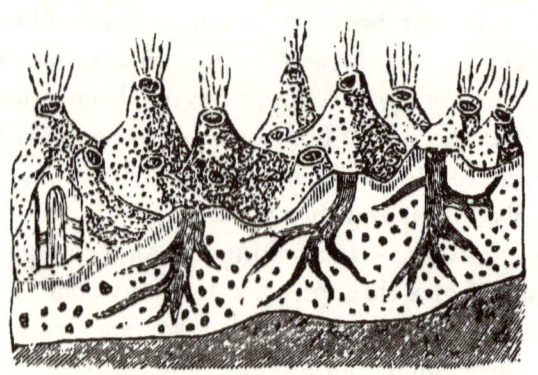

Fig. 5.—LIVING SPONGE IN ACTION.

lashes, because they are, in fact, formed of soft, delicate flesh.   It is their business to wave gently but continually, and thus urge onward the flowing current of water.   Notic-

ing the arrows, we may now follow the direction of the
tiny streams as they enter the small pores, pass through the
canals, and are finally thrown out from the large holes on
the surface. With a microscope little fountains like those
represented in Fig. 5 may
be seen constantly play-
ing from the large holes
of a living sponge.

8. **How Sponges Eat.—**
Everything that lives
must eat and breathe,
but how is the sponge to
eat without a mouth?
When the food touches
any part of its body, the
soft, jelly-like flesh sinks
in to form a little bag, and
at the same time the sur-
rounding parts creep out
over the morsel of food
until it is entirely covered
and digested. After this
the flesh returns to its
original position, and any
shell or other refuse that
remains from the meal is
washed away.

9. **The Young.—**Sponges
have a curious manner of
producing their young.

Fig. 6.—Neptune's Glove.

At certain seasons very small
oval masses of jelly are formed on the inner surface of
the canals, which finally drop off. They remain in the
canals for a time, and become perfect eggs, after which

they are thrown out by the water forming the little fountains, and instead of falling to the bottom, as we might
suppose such helpless masses of jelly would do, they swim
around as if they meant to have a little sport before commencing the sober realities of life.

10. **Food for other Animals.**—You will be interested to
know that while these jelly-like eggs were resting in the
canals of the parent sponge, delicate cilia (which we learned about just now) were forming at one end of the egg.
These cilia strike the water with a rapid motion, and the
eggs are rowed about through it until they settle down
and attach themselves to some rock or shell on the bottom of the ocean, and finally grow up into the perfect
sponge. The waters are swarming with these eggs at
certain seasons, and great quantities of them are eaten by
larger animals.

11. **Size and Shape.**—Sponges are common in nearly all
parts of the world, and they differ greatly in size and
quality, but only a few species are useful to man. Some
species are nearly round, others are always cup-shaped,
some top-shaped, and some branched. A fresh-water
sponge is frequently found in our streams, growing upon
sticks and stones. It is of a bright green, and when
seen under the water in the sunlight it is very pretty.

12. **Spicules of Sponges.**—The spicules of sponges grow
in a variety of elegant shapes; generally they are visible
only with a microscope. They are composed of lime or
flint, and are usually sharp-pointed. They are embedded
in the flesh as well as in the horny fibres, thus serving
to protect the helpless creatures from being devoured by
fish and other animals. In our fine sponges the skeleton
is almost destitute of spicules, while in some others these
spicules are very numerous and wholly support the flesh.

Fig. 7.—VENUS'S FLOWER-BASKET.

Fig. 8.—SPONGE-FISHING.

Such sponges are so loose in texture that they are of no value for domestic purposes.

13. **Where Found.** — Fine sponges are used by physicians in surgical operations, and are sometimes very expensive. Our finest sponges come from the Mediterranean Sea and the Red Sea. They are obtained by divers, who search for them under rocks and cliffs, and who remove them carefully with a knife, that they may not be injured. The Turks, who carry on the trade, have between four and five thousand men employed in collecting sponges, and the value of those collected each year is estimated at ninety thousand dollars. Coarse varieties are found in the Gulf of Mexico and the Bahama Islands. They are scraped off the rocks with forked instruments, and consequently.they are often torn.

14. **Method of Culture.** — The demand for sponges has

1*

increased so much during the last few years that there is cause to fear the supply will be exhausted, unless some way can be found to cultivate them by artificial means. With this view, attempts have recently been made to raise sponges in the Adriatic Sea and in the Gulf of Mexico, by taking cuttings from full-grown ones and fastening them upon stones on the bottom of the ocean until they attach themselves. These experiments have been successful, but the operation is a delicate one, requiring great care not to bruise the soft flesh. It is necessary to keep the sponge under sea-water during the process.

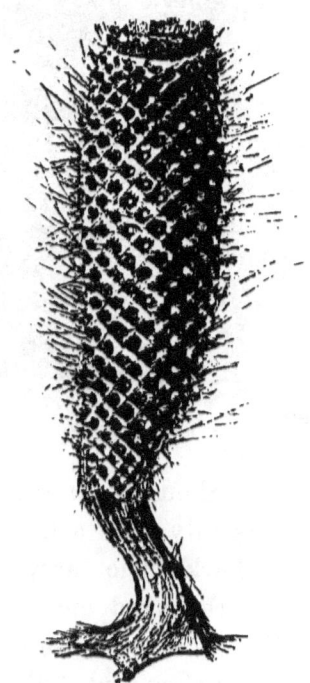

15. **Glass Sponges.**—Some of the glass sponges are exceedingly beautiful. One of these, the delicate "Venus's flower-basket," grows in the deep sea near the Philippine Islands. It looks like spun-glass woven into a beautiful pattern, and is so exquisite we can scarcely believe that it is the skeleton of a sponge. Fig. 9 shows another variety of glass sponge found between Gibraltar and the island of Madeira by the scientific party on board the famous *Challenger*, which ship was sent out by the British Government to explore the animal and vegetable wonders of the great deep.

Fig. 9.—GLASS SPONGE.

16. **Boring Sponges.**—The "boring sponge" spreads itself over the shells of oysters and mussels, boring them

through and through, and dissolving the shell. It even bores into solid marble, and will, in time, completely destroy it.

17. Flints are exceedingly hard substances, yet they are supposed to have been formed from soft sponges. By examining small pieces of flint under a microscope the texture of the sponge, in a fossil condition, is often clearly seen, and the spicules peculiar to sponges are recognized.

II.

HYDROIDS.

SOME ODD RELATIONS OF THE JELLY-FISHES.

SUB-KINGDOM, CŒLENTERATA : CLASS, HYDROZOA.

1. Hydroids, or Sea-firs.—Let us now examine some odd-looking animals called "hydroids," or sea-firs, which grow in the ocean, firmly rooted upon the bottom, or attached to shells and stones. The tall branches in Fig. 10 are hydroids growing upon the shell of a dead mussel. A barnacle, too, has lived and died on this pretty shell, and little sea-weeds cluster around its remains.

Fig. 10.—HYDROIDS GROWING ON A SHELL.

2. Related to Jelly-fishes.—We can scarcely imagine animals that are more unlike jelly-fishes than these slender branching hydroids are; and yet the wonderful story I have to tell you will show them to be so closely related that we could not study the life of one without studying the life of the other.

3. Their Resemblance to Plants.—Long graceful sprays

of hydroids are often thrown on shore by the tide, and as they resemble plants much more than animals, they are generally mistaken for sea-weeds. Many persons gather them for decorating brackets and hanging-baskets, and we frequently see bunches of them arranged in sea-shells, and offered for sale in our shops. The shopkeeper would probably not know them by any other name than sea-weed. Still, they are animals, and we can mostly recognize them by their yellow, horny appearance, and by the numerous joints on their stems.

4. **Each Spray-point bears a Cup.**—In looking at one of these sprays with a microscope you will find each little point on the stem to be in reality a dainty cup, which when alive contained a hungry animal. Should you find a piece freshly washed up from the ocean, it would be well to place it in a glass jar filled with sea-water, and after allowing it to remain perfectly still for a while, it may perhaps show you, if it is yet alive, how it has been accustomed to pass the quiet hours in its native home.

5. **Hydroids higher in the Scale of Life than Sponges.**— You will find each cup occupied by a soft animal, with a mouth in the centre opening directly into the stomach. Hydroids, you see, are higher in the scale of life than sponges, for they possess mouths and stomachs. As we watch, the body of the animal will rise up in the cup, and from around the mouth will gradually creep out slender thread-like feelers, which may be extended quite a distance, or drawn up at will entirely within the body of the animal. You will, of course, wish to use the proper name for these feelers. They are called tentacles, and they evidently serve to produce currents of water towards the mouth, and to bring the required food. In this way the little animals live, day after day and year after year, pa-

tiently waving their tentacles, and waiting for the food that is sure to come.

6. **How Hydroids produce Jelly-fishes.**—Do you ask what connection there is between these quiet little animals and the active jelly-fishes? We shall soon see. The hydroids have grown by budding and branching somewhat as plants do. Occasionally pear - shaped cups much larger than those we have looked at are formed on the stem. These

large cups are called spore - sacs. They contain the substances which, later, will grow into eggs, and at the proper time they fall off. After resting a while, and throwing out cilia and tentacles, these spore-sacs swim gayly away, and, strange to relate, they are henceforth known by the name of jelly-fishes!

7. **The Spore-sacs.**—In Fig. 11 you will see a spray of hydroid magnified which shows two spore-sacs. In the species which is represented here (the *Sertularia*) the spore-sacs do not fall off, but they burst and discharge the eggs which they contain. These jelly-fishes now lead active lives, and as they dart and swim about in the

Fig. 11.—HYDROID MAG-NIFIED, SHOWING SPORE-SACS.

water no one would suspect that they had any relation to the plant-like animals with which we started, yet it is supposed that most hydroids have this wonderful history.

8. **The Young unlike their Parents.**—Jelly-fishes produce eggs, from which are born little floating bodies. These after a time fasten themselves to some stick or stone, and grow by budding until they become the elegant feathery

branches which we must now call hydroids. The young of nearly all animals resemble their parents, but the children of jelly-fishes, you see, are very different from the jelly-fish itself. In the next generation, however, we shall find jelly-fishes again.

9. **Difference between Plants and low Forms of Animal Life.**—Most of the plant-like objects which we are accustomed to see growing near the shore are in reality hydroids. Has it ever puzzled you to know the difference between plants and these low forms of animal life? One very important difference is that most plants can procure their food directly from the soil, whereas animals are obliged to feed upon living substances, or those which have at some time been alive, as vegetables and animals.

10. **Found in all Parts of the Ocean growing in Families.** —Hydroids grow in all parts of the ocean, in deep water as well as near the coast. Some of them are three feet high. One branch may contain a hundred thousand distinct animals, the only communication between them being a circulation of fluid through the hollow stems. In this way each branch constitutes a family which has sprung originally from the same little egg. Some varieties never grow tall, and as they occur in patches over rocks and shells, they resemble thick beds of moss.

11. **Another Manner of producing Jelly-fishes.**—The little hydroids which we see hanging from the under side of a rock in Fig. 12 produce jelly-fishes in a different manner from the one I have described, although it is equally remarkable. This hydroid has no buds or branches, but the main tube of the body divides itself into a number of rings or plates, until the whole animal looks somewhat like a pile of tiny saucers with scalloped edges. Finally the upper plate begins to twist and squirm until it loosens

itself from the pile, and floats off to lead the gay and independent life of a jelly-fish. It is followed by the other plates in their turn, each making a separate animal. These new jelly-fishes eat greedily and grow fast, forming some of our largest varieties.

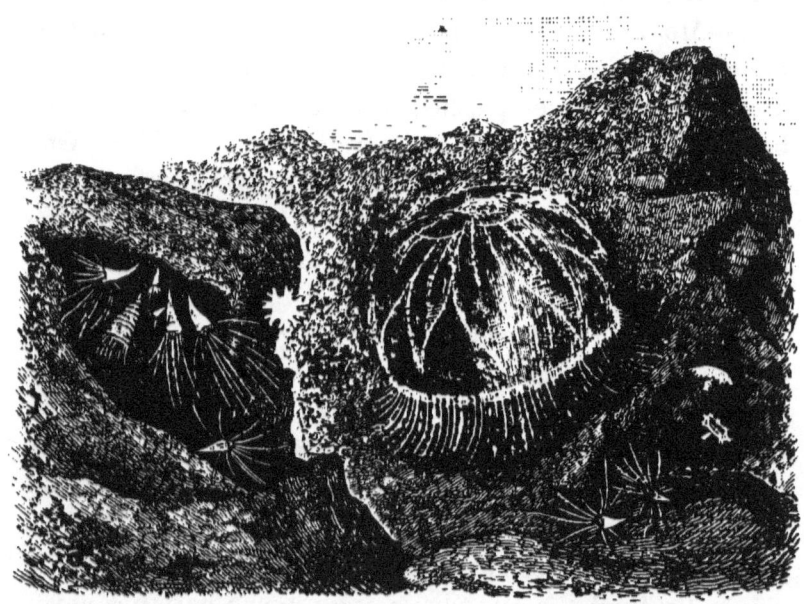

Fig. 12.—Jelly-fish (Aurelia Aurita), with Young in Various Stages.

12. We can form but little idea of the immense numbers of animals living in the ocean and drawing from the surrounding water all that is needed for their support. They cannot go in search of food, and they take only such as floats towards them. Still, they seem to have some choice in the matter, as they reject from their mouths any food they are not suited with. Many of these curious animals have bright colors, and surrounded as they are with a great variety of plants, they give to the bottom of the ocean a marvellous beauty.

13. Does it not seem strange that the slender, delicate sprays of which we have been speaking are really animals, and more than that, the children of jelly-fishes ? A little girl once exclaimed, on hearing of these wonderful changes that happen in the life of hydroids, " Why, it seems almost like a fairy-tale !"

III.
JELLY-FISHES.

SUB-KINGDOM, CŒLENTERATA: CLASS, HYDROZOA.

1.· **Jelly-fishes.** — When jelly-fishes are seen lying in shapeless masses upon the beach, where they have been washed by the tide, their appearance is not attractive. If, however, we can watch them from the side of a boat,

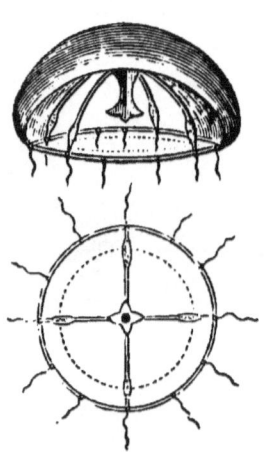

or from a long pier, as they dart through the water with their tentacles trailing after them, we shall soon learn to admire their graceful movements and their elegant colors. There is something very interesting too in these little inhabitants of the great deep. They are such soft, helpless things, yet they live and have their own good times if only the boisterous waves do not catch them and dash them too harshly against the rough shore.

Fig. 13.—SECTION OF JELLY-FISH, SHOWING TUBES AND MOUTH.

2. **Jelly-fishes a single Bell-shaped Mass.**—Jelly-fishes consist of a single bell-shaped mass of jelly, from the inner surface of which hangs the body of the animal, with the mouth in the centre. The mouth opens directly into the stomach, from which several hollow tubes (usually four) extend to a circular tube around the edge of the bell. In the jelly-

fish (Fig. 13, *a*), the side next to us has been removed, that we may see the tubes and the mouth hanging in the centre; *b* shows us the same viewed from below.

3. **Eggs of Jelly-fishes.**—The eggs of jelly-fishes are formed in large quantities in the tubes leading from the centre. Fig. 13 shows the enlarged cavities containing eggs. At certain seasons of the year great clusters of bright-colored eggs may be seen through the transparent flesh. A few jelly-fishes are thought to produce young ones resembling themselves, without passing through the strange forms we noticed in studying hydroids.

4. **How they Move.** —Hydroids, you will remember, are

Fig. 14.—JELLY-FISH (CYANŒA ENPLOCAMIA)

abundant in all oceans; so are jelly-fishes, and they are often found floating in large companies. Jelly-fishes are propelled by alternately taking in and throwing out water under the bell. This gives them a jerking movement, as though caused by breathing. They come to the surface chiefly when the water is quiet, and, as they like the warm sun, you will not see many of them at an early hour in the day. They are easily alarmed. If they meet with an obstacle in their course, or if they are touched by an enemy, the bell contracts, the tentacles are instantly drawn up, and the creature sinks in the water.

5. **Beginnings of Eyes and Ears.**—Upon the outer edge of the bell there are bright-colored specks and solid spots, which are thought to be the beginnings of eyes and ears. Although these spots never grow to be perfect eyes and ears in the jelly-fish, they promise that Nature has in store for her children the precious gifts of sight and hearing.

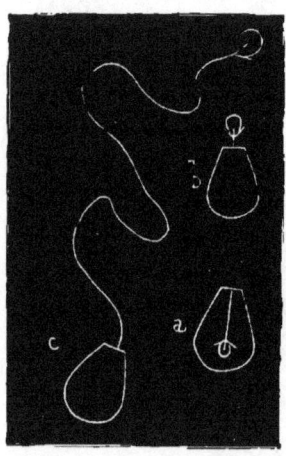

Fig. 15.—LASSO-CELLS FROM A FRESH-WATER HYDROID (MAGNIFIED).

a, Barbed Dart within the cell; *b,* Barbed Dart escaped from the cell; *c,* Lasso fully extended, carrying the dart at the end.

Such imperfect organs are called by the wise men rudimentary organs. This is the lowest animal in which anything corresponding to our nerves is found.

6. **Power of Contracting and Expanding their Tentacles.**—Delicate fringes and tentacles hang from the lower edge of the bell, adding greatly to its beauty. The tentacles are often many feet long, yet the animal has the power of drawing them up so that they are not visible. This curious power of contracting and expanding the tentacles belongs to many humble sea creatures. Sometimes, while we are wondering at their disappearance, they lengthen again as if by magic.

7. **How Jelly-fishes secure Food.**—The tentacles of jelly-fishes are covered with a great many lasso-cells. These

lasso-cells are too small to be seen without a microscope; still, they are powerful weapons in their way, and are quite sufficient to enable the jelly-fish to catch its food. You know how the skilful hunter uses a lasso for catching wild cattle. The jelly-fish uses its lasso in quite a different manner, but it may be equally effective.

8. **Lasso-cell.**—When examined, each lasso-cell, or little sac, is found to contain a long slender thread coiled within it, somewhat like a lasso, and floating in a fluid. The cell is filled so full of the fluid that it bursts with the slightest touch, and as the fluid squirts out, it carries with it the slender lasso armed with sharp stings. In this way lassos are darted out to capture many little crabs or fishes that brush too near in passing.

9. **Description and Use of Lasso.**—The sting of the lasso seems to paralyze the unfortunate creatures, and they make no effort to escape as the tentacles coil round them and carry them to the mouth of the greedy jelly-fish. In Fig. 15 you will see a group of lasso-cells highly magnified. The cell at *a* has not yet burst, and through its thin walls we see the barbed dart at the end of the lasso. At *b* the lasso has been thrown out only a short distance, while at *c* the long slender lasso still carries the dart at the end, and the curious little bladder is much larger than it was inside the cell. The lasso of this specimen is exceedingly delicate and simple, while that of some animals is covered with stinging bristles. Is it not a dainty weapon to be used in the continual warfare carried on by these innocent-looking creatures? Small as the lassos are, they serve to protect the soft-bodied animals from their numerous enemies.

10. **Sea-nettles—Medusæ.**—If we should touch the soft,

pretty tentacles of a jelly-fish, we should probably be stung by these tiny weapons. This irritation is produced in the flesh by the numerous sharp points on the lassos, and is similar to the stinging of nettles. For this reason jelly-fishes are often called sea - nettles. The correct name, however, which you will find in scientific books, is "Medusæ."

11. **Size.**—Jelly-fishes vary greatly in size. Some are mere dots, so extremely small that we might not notice them in the water, while one species is said to be seven feet in diameter, with tentacles measuring fifty feet in length. The parent of this huge jelly-fish was a hydroid only half an inch high. Its children will be the same. What do you think its grandchildren will be?

12. **Nature of the Jelly-like Flesh.**—The jelly-like flesh of these animals consists largely of water, and a specimen weighing several pounds when alive will shrink away to almost nothing if exposed to the sun and the wind. As the body contains no bones or other solid matter, it all perishes together, and no trace is left of its former beautiful shape. You will see that jelly-fishes are in no way like real fishes. As they float on the ocean they look more like fantastic mushrooms, and one writer has called them "Mushrooms of the Sea."

13. **Color.**—It would be impossible to describe the varied colors of jelly - fishes, as they include almost every hue, the beautiful tints being probably heightened by their transparency. All shades are to be found, from pale blue and pink to bright red and yellow. Those found in tropical seas are of a deeper color than ours.

14. **One Delicate Kind.**—In striking contrast with these brilliant jelly-fishes is one species which is so delicate and transparent that as it floats upon the water we can scarce-

ly see the sub-
stance of which
it is composed.
The only parts that
strike the eye are
the circular tube around
the edge and the four ra-
diating tubes with their
large clusters of eggs.

Fig. 16.—MUSHROOMS OF THE SEA.

The tubes look as if they were held together by some
slight web. This jelly-fish is extremely languid in its

movements, and it sometimes remains perfectly quiet in the bright sunshine for hours, not even moving its tentacles.

15. **Phosphorescence.**—Although jelly-fishes are so brilliant in the daytime, they have a different beauty at night, when they throw out a golden light slightly tinged with green, resembling that of a glowworm. Vast numbers of small animals in the sea have this power of giving light from their bodies. The light is called phosphores-

Fig. 17.—GROUP OF PHOSPHORESCENT ANIMALS.

cence. As it may be seen at any time of the year illuminating all oceans, it is an unfailing source of delight to voyagers. It is most conspicuous on a dark night, when the water is agitated by the motion of a boat, or by the breaking waves. In Fig. 17 is a group of the larger phosphorescent animals.

16. A pail of sea-water carried into a dark room often affords a good opportunity for studying this interesting phenomenon. Although we may not have detected the presence of any animals before, when the water is stirred or jostled we may see the beautiful sparkles of light. The phosphorescence of some animals is of a bluish tint; in others it is red, like flame.

17. A person will rarely tire of watching a boat as its prow turns up a furrow of liquid fire, and each dip of the oar sends a miniature flash of lightning through the otherwise dark water.

2

IV.

THE "PORTUGUESE MAN-OF-WAR."

SUB-KINGDOM, CŒLENTERATA: CLASS, HYDROZOA.

1. SOME one has probably imagined that this curious floating animal looks like a Portuguese war vessel, and on that account has given to the innocent and defenceless creature a name which seems to us very inappropriate. We will not be dismayed, however, by a forbidding name, for the graceful animal is not in the least warlike. It is to be hoped you may all have the pleasure some day of seeing one floating over the sea like a fairy vessel, not minding the winds or the storms. You will be delighted with its beauty, and you will wonder how so frail a bark can withstand the waves.

2. **Shape and Color.**—When we examine the Portuguese Man-of-war we shall find it to be a transparent pear-shaped bladder, about nine inches long, throwing off, like a soap-bubble, bright blue colors tinged with green, violet, and crimson. On top of the bladder there is a wavy, crumpled crest of delicate pink. This may perhaps act as a sail.

3. **A Colony of Animals.**—From one end of the bottom hangs a large bunch of curious-looking, bright-colored threads, and bags, and coiled tentacles which trail after it. You will see these streamers in the picture, and you may be surprised to learn that they are separate animals, forming a little colony, which is floated by the bladder.

Still, they are not entirely distinct; they have various uses, and each contributes its share to the good of the colony. Some produce eggs, some do the swimming, some do the eating, and others are provided with lasso-cells to procure food.

4. **The Food taken by One nourishes All.**—In such colonies of animals as this, the food which is taken by one individual helps to nourish all the others. This is accomplished by the circulation of fluids throughout the whole colony, carrying nourishment to each one.

5. **Organs Defined.**—In animals that are more highly developed we shall find these offices performed by special parts of the same body. These different portions of the body, which are set apart to perform certain duties, are called organs. Thus we speak of the eye as the organ of sight, and the ear as the organ of hearing.

6. **Length of Tentacles.**—The tentacles of the Portuguese Man-of-war are more than twenty feet long, yet they may be drawn up to within an inch of the bladder. The lasso-cells upon their surface not only wound the prey, but also sting bathers or any persons who come in contact with them. Even after death the tentacles produce irritation when they are touched.

7. **Where Found.**—These beautiful creatures are found in tropical seas. They are abundant in the Gulf of Mexico, and are often carried by the Gulf Stream into Northern waters. Occasionally they drift upon our own shore. Do you think you would recognize one floating on the ocean when you had not expected to see it? If you should ever have one in your possession, it may be well to dry it or keep it in alcohol; for although its delicate beauty cannot be preserved, it will still be interesting to those who have never seen living ones.

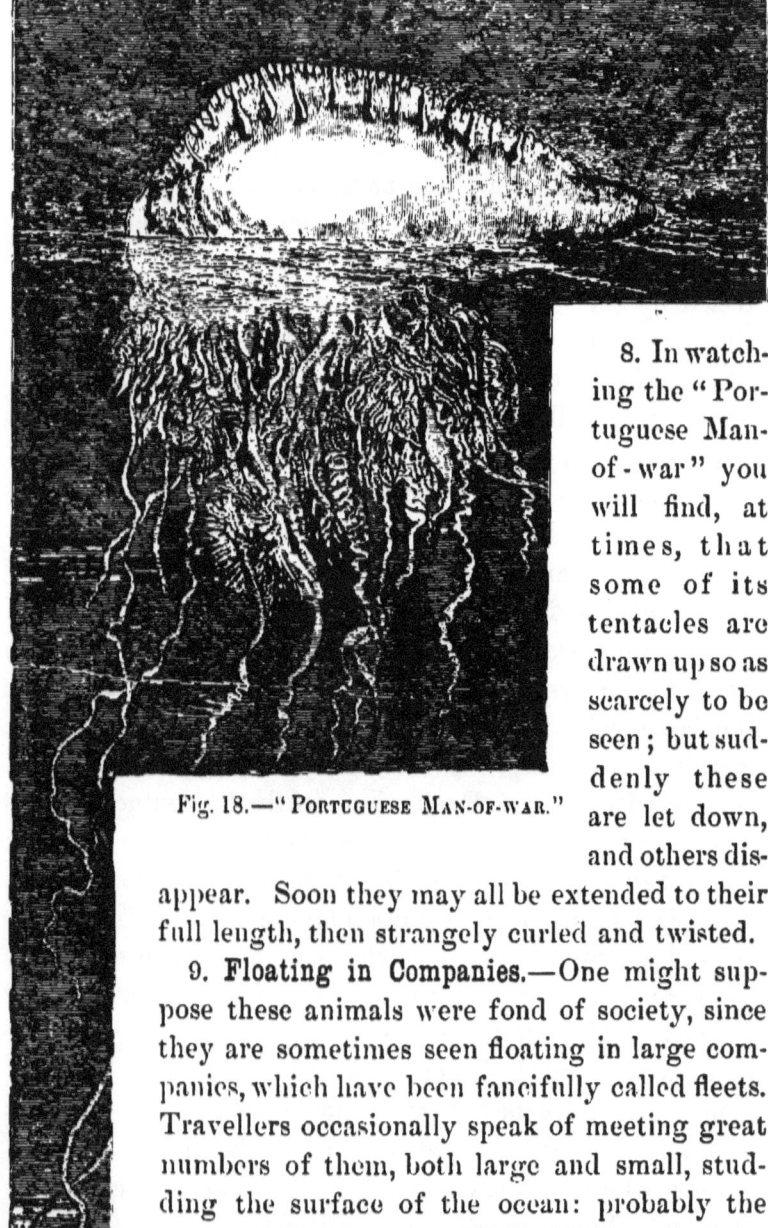

Fig. 18.—" PORTUGUESE MAN-OF-WAR."

8. In watching the "Portuguese Man-of-war" you will find, at times, that some of its tentacles are drawn up so as scarcely to be seen; but suddenly these are let down, and others disappear. Soon they may all be extended to their full length, then strangely curled and twisted.

9. **Floating in Companies.**—One might suppose these animals were fond of society, since they are sometimes seen floating in large companies, which have been fancifully called fleets. Travellers occasionally speak of meeting great numbers of them, both large and small, studding the surface of the ocean: probably the young ones were out sailing with their parents.

V.

SEA-ANEMONES.

SUB-KINGDOM, CŒLENTERATA: CLASS, ACTINOZOA.

1. **Ocean Treasures.** — Many of you, no doubt, have learned, when at the sea-shore, the delight of climbing over wet rocks covered with slippery sea-weed, and peering into the little pools left between the stones to see if the great waves have dropped any treasures from the ocean. Those who have enjoyed this pleasure will gladly recall the sparkling pools, carpeted with rich-colored sea-weeds which half conceal the timid animals that live there.

2. In such pools the rocks, the shells, and the sea-weeds all have richer tints from the bright water that covers them, and one who loves beautiful things will linger beside the pools as if gazing into enchanted gardens.

3. **Sea-anemone.** — On searching these rock pools we shall probably find many curious animals. None would interest us more than the "sea-anemone," though when we find it hiding in some dark corner, with its tentacles all drawn in, and looking like a soft lump, it may not promise much beauty.

4. **Why so Named.** — The sea-anemone adheres firmly to the rocks, so we will not pull it off. If we watch long enough we shall see it begin to rise in the middle, and from the summit will creep out, very slowly and softly,

beautiful tentacles like a wreath around the top. It is now that this singular animal looks like a flower, and deserves the name that it possesses. Perhaps it is not so much like an anemone as it is like a chrysanthemum or some other flower with a great many petals. You will be charmed with the delicate light-colored tentacles waving gently in the water.

5. **Description.** — In the middle of the tentacles is the mouth, leading into a hollow sac, which is the stomach.

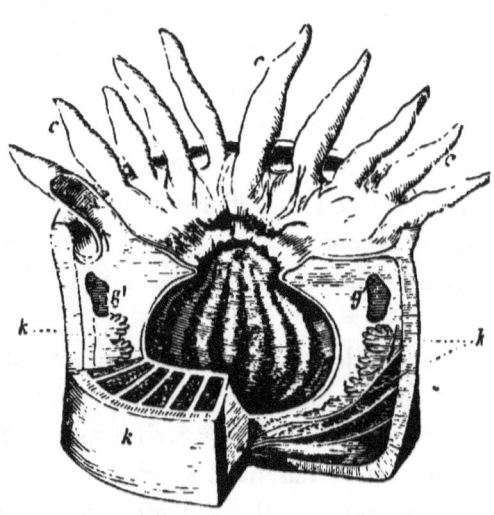

The remainder of the body is divided by partitions from top to bottom into open chambers. In Fig. 19 you will see the stomach at *e*, and the chambers at *k*. There is an opening at the bottom of the stomach through which the food passes after it is digested. Sea-water also enters the body through the stomach, and both the

Fig. 19.—STOMACH AND CHAMBERS OF SEA-ANEMONE.

c, tentacles; *d*, mouth; *e*, stomach; *g'*, *g'*, openings in the partitions; *k*, chambers.

water and the nourishment circulate freely through the chambers. Each tentacle is a hollow tube connected at its base with one of the chambers, and readily filled with water. Here we have an explanation of the mysterious manner in which the sea-anemone swells itself out and then shrinks away again. The body and tentacles are enlarged by drawing in water to fill them, and when they suddenly contract the water is forced out through the mouth.

6. **No hard Skeleton.**—The sea-anemone has no hard skeleton whatever; all parts of the body are soft, like a stiff jelly. It can draw in its tentacles out of sight, and it will do so upon the slightest alarm, rolling itself into an ugly lump like the one we found. Allow it to remain quiet for a while, however, and it will blossom out as gorgeously as ever.

7. **The Manner of Feeding.**—When any little crab, or worm, or small fish brushes past the tentacles, the lasso-cells are darted out to paralyze it, and the tentacles then seize the prey and carry it to the mouth. The bones or shells which remain after the meal are thrown out from the mouth. The tentacles hold the prey tightly, so that even cunning crabs cannot escape, and you know it is not the easiest thing in the world to catch a crab and hold it.

8. Sea-anemones are greedy creatures. It takes a great deal of food to satisfy their appetites, and their mouths can be extended to receive quite large animals. They eat mussels and cockles by sucking the body out of its shell. Great numbers of sea-anemones, in their turn, are devoured by other animals, their soft bodies offering little resistance to their foes.

9. **Variety of Color.**—The variety of color in these animals is almost endless. Some of them are rich orange and chocolate colors, others purple dotted with green. One beautiful species has violet tentacles pointed with white; another, red tentacles speckled with gray. Another spreads out its green arms edged with a circle of dead white, while still another opens a milk-white top circled with a border of pink. In Fig. 20 is a cluster of beauti-

ful anemones. The two small ones at the right show how these creatures look when closed.

10. **Animals protected by "Mimicry."**—Some sea-anemones which live in exposed situations are of a dull, dusky brown, covered with rough warts, while animals of the

Fig. 20.—Cluster of Anemones.

same species, living in deep water, where there is less need of concealment, have smooth skins adorned with brilliant tints of rose, scarlet, or light green. This beautiful provision of Nature for protecting animals by making them inconspicuous is called "mimicry." In following our studies in Natural History we shall find many instances of this general resemblance in the color or shape of animals to the objects by which they are surrounded, and we shall notice that the animals are in this way concealed from their enemies.

11. **Great Numbers of Eggs.**—The number of eggs produced by sea-anemones seems almost incredible. A single animal is said to throw out three hundred eggs in one day. The eggs are little jelly-like lumps which are formed on the inside of the partitions, and are thrown out from the mouth. After swimming about by means of cilia, they settle on some solid body and begin to grow. Sometimes the young ones remain within the body of the parent until their tentacles are formed. They are then ready to settle down soon after reaching the water.

12. **Budding and Renewal of Lost Parts.**—Sea-anemones increase by budding as well as by eggs. At the lower edge of the body little round knobs are sometimes formed, which separate from the parent and grow into perfect animals. If the tentacles or other parts of the body are removed, new tentacles soon grow in their places. If an individual is torn in pieces, each fragment has the power of forming for itself a mouth and throwing out tentacles, and becoming a new sea-anemone, perfect in all its parts.

13. **Where Anemones are Found.**—Most species live in holes among the rocks, attached to stones or shells, over which they slide in a clumsy way. They are especially fond of deep dark grottos, and when they have taken full possession of such places, they may be found clinging to the sides and roof of the cave, and displaying their charms without reserve. Although they do not enjoy the glare of the bright sun, they expand best in mild, clear weather, and remain closed when the sea is rough and stormy.

2*

Fig. 21.—Sea-anemones.

14. A few of these animals float on the ocean. One sea-anemone is fond of a roving life, and having no very good means of travelling about, it attaches itself to the back of a certain kind of crab, and accompanies the crab

in all its wanderings. There seems to be an attraction between the two, and one is rarely seen without the other.

15. Another species is mostly found clinging to the shell of a whelk, but for certain good reasons it never clings to a living one. The whelk burrows in the sand. This would be disagreeable and inconvenient to the anemone, so it prefers a dead shell which has been taken possession of by a hermit-crab, and henceforth it travels about with the crab. We should scarcely look for affection in a crab, but it has been said that the hermit grows fond of its companion, and that when it has outgrown its shell and has selected a new one, it will carefully lift the anemone from the old home and place it on the new one, " giving it several little taps with its big claws to settle it."

16. **A Simple Aquarium.**—Do not fail to hunt up these lovely rock pools when you have an opportunity. The pleasure of a visit to the sea-shore is greatly increased by an interest in the strange forms of animal life which we see there and nowhere else. A glass jar filled with sea-water is often a source of great delight. In it you may drop any strange-looking object that has excited your curiosity. Perhaps this very object may prove to be some odd little animal which is not yet dead, but which will revive with the touch of the life-giving water.

17. In this way we may watch their habits and their hidden beauties. Sea-anemones do well in such an aquarium, and as they cling to the side of the jar, we can observe all their parts while they are in action. By far the pleasantest way to learn about them is to let them tell their own story. The water must be changed frequently, for impurities are constantly passing from the bodies of even these delicate animals. They will soon die if placed in fresh water.

VI.

CORALS.

SUB-KINGDOM, CŒLENTERATA: CLASS, ACTINOZOA.

1. **Corals.**—Most persons admire corals. They are so common and easily obtained that I hope each of you will lay aside your reading just here, and hunt up a piece, no matter how small, that we may examine it carefully, and see what we can find out about it. You must find, however, a piece of the natural coral, just as it was brought up out of the sea, and not a polished piece such as is made into ear-rings and brooches and strings of beads for ornaments.

2. **The Roughness on the Surface.**—What makes this bit of natural coral so rough? The first glance will convince you that those curious pits and little cups on the surface mean something; and when we remember that all the corals which reach us are the skeletons of former living animals, they interest us at once.

3. **Home of the Corals.**—Few of us, perhaps, will ever be so fortunate as to see living corals, since they grow principally in the deep water of warm oceans. The higher the temperature, the greater the variety and profusion of corals. They are delicate creatures, however, as they will not flourish under adverse circumstances. They require water of a certain depth, and they die immediately if exposed to the sun or to cold weather. During life the skeleton is covered with soft flesh, the surface of which is thickly studded with star-like animals called polyps. In

Fig. 22.—BRANCHING CORAL ALIVE, WITH POLYPS EXPANDED.

this way millions of polyps are sometimes clustered together in one community. As they wave their delicate
tentacles of white, green, or rose color, they are very beautiful, especially if seen in bright sunlight through water
that is clear and still.

4. **A Piece of Coral Examined.**—In Fig. 22 is shown a
piece of living coral with the polyps expanded. The flesh
has been removed from the upper branch on the left that
we may see the skeleton. Let us suppose that the specimens we have selected for study are of this kind. Each
of the tiny cups on the surface was once the framework

of a separate polyp, and we shall find that its interior is divided by a number of partitions which do not quite reach the centre. Look into the cups with your microscopes,* and you will find them very beautiful. One set of partition-walls reaches almost to the centre, and between these walls are shorter ones. These give us a clew to the kind of animal that has lived here, and they will at once remind you of the partitions in the sea-anemone, as shown in Fig. 19 of the last chapter. Indeed, the whole structure of a coral polyp is similar to that of an anemone, and we can now easily imagine the stomach of the polyp hanging down in the opening left between those delicate partitions. Coral polyps differ from sea-anemones, however, in three important ways—they have hard skeletons, they cannot move about, and they usually grow in clusters.

5. **Life History of the Polyps.**—When young, coral polyps are little jelly-like animals which swim about in the water. After they have chosen a resting-place, and the stomach and tentacles have grown, hard particles of lime, which they have drawn in from the sea-water, settle in their flesh to form a circular cup as well as the partitions inside. In this way the polyps soon acquire a solid frame, the soft parts being the stomach, the fringe of tentacles, and the fleshy mass covering the skeleton. They can draw the tentacles entirely within the body, as the anemone does. Like the anemone they also have lasso-cells for capturing their food.

6. **How Corals become Branched.**—Should it be a branching coral whose history we are tracing, it will now begin

---

* A Coddington lens, which is inexpensive, is a useful thing to possess. It can be carried in the pocket; and if we have it always with us, we may find new beauties wherever we go.

to bud from the sides. The buds will grow into branches, throwing out other buds, somewhat as plants do, until we have an elegantly branching colony. Each bud is a new polyp, and remains attached to the branch from which it sprang. Although the polyps in such a community have separate mouths and stomachs, there is a close connection between them, and a free circulation of fluids through the soft flesh.

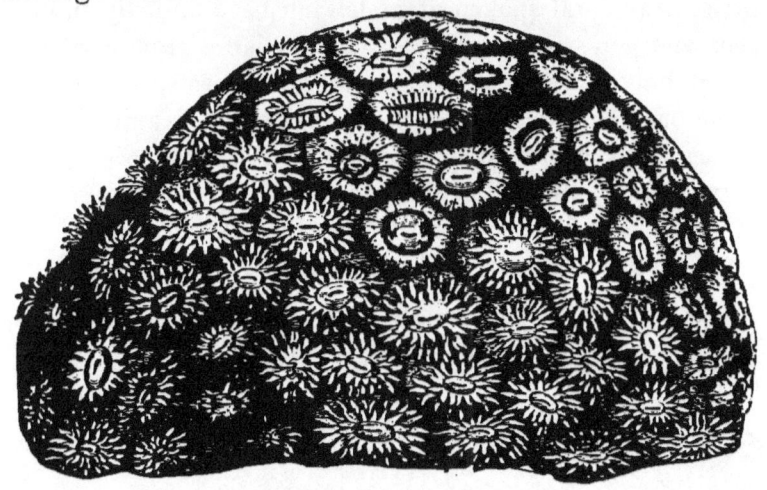

Fig. 23.—ASTRÆA PALLIDA (NATURAL SIZE).

7. **One Generation after Another.**—As in other families one generation passes away and another takes its place, so in large branches of coral the lower and older portions may be dead, and living polyps will be found only at the ends of the branches.

8. **The Eggs.**—Besides increasing by budding, corals increase rapidly by eggs. Their eggs are pear-shaped, transparent bodies, covered with cilia, which are in constant motion, and which row the jelly-like lumps through the water. The parents, you remember, are firmly rooted to some object, but their little ones are gifted for a time

with the power of motion. They may well enjoy the privilege while it lasts, for it is their only chance of exploring their ocean home. Presently they must settle down like other sedate corals. It is in this manner that the young polyps are distributed through the ocean instead of growing in a crowded colony around the parent.

9. **Coral not Built by an Insect—"Radiates."**—You will often hear coral spoken of as having been built by an insect, and you will see at once that this is far from correct. Coral polyps are very different from insects, and their skeletons grow, much as ours do, inside of the animal; so we cannot say they have been built. All such animals as coral polyps, which have the mouth in the centre, with other parts radiating from it, are called "Radiates."

10. **Different Forms of Coral.**—Besides these branching

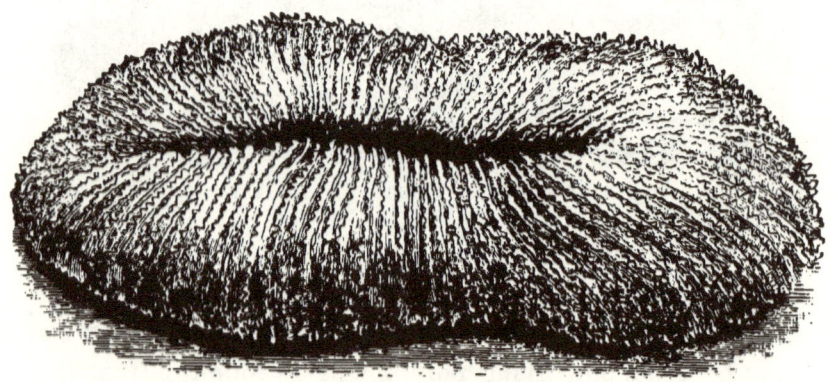

Fig. 24.—MUSHROOM CORAL.

corals which resemble trees and shrubs, there are other kinds which grow in solid masses without sending off branches. Some assume the shape of graceful vases, and all of these varieties are gayly decked with star-like polyps of varied colors. Does it not seem to you as if the ocean was one vast storehouse of beautiful things?

11. **The Mushroom Coral.** — The mushroom coral (Fig. 24) looks indeed like a large mushroom, although you will notice that the leaflets are on the upper surface instead of being underneath, as they are in the vegetable mushroom. This coral is the skeleton of one huge polyp, and we see the depression in the centre corresponding to the little cups on most other corals.

12. **Organ - pipe Coral.** — The organ-pipe coral consists of lovely crimson tubes standing upright, and connected at short distances by thin flat plates, which give it the appearance of being several stories in height. These plates may be distinctly seen in Fig. 25. When alive, a bright purple polyp protrudes from the top of each tube.

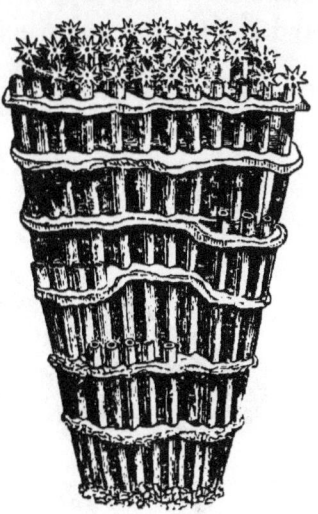

Fig. 25.—Organ-pipe Coral.

13. **Red Coral.**—Red coral, which is used for jewellery, grows in a bushy form on rocks at the bottom of the Mediterranean and Red seas. The fleshy mass of this coral is colored red by the numerous red spicules it contains, while the polyps themselves are pure white, the whole resembling a pretty red shrub spotted over with sparkling white flowers. The spicules in the centre of the branches form a solid stem, which takes a fine polish. Underneath the flesh the surface of the coral is marked with deep grooves, which are canals for the circulation of water. These grooves are shown at both ends of the

branch in Fig. 26. They are always removed in polishing.

14. Red coral is generally obtained by fishermen, who drop into the water heavy wooden crosses to which strong nets are attached. As the boat moves slowly forward, the crosses are raised and lowered to break off the coral branches. The apparatus is then lifted from the water, and the fragments of coral which have become entangled

Fig. 26.—FRAGMENT OF RED CORAL WITH POLYPS.

in the net are carefully removed. There are many shops in Italy where the coral is polished and cut into various ornaments. Delicate rose-colored corals are considered very choice and elegant, but the natives of India prefer blood-red ones, which contrast finely with their dark rich complexions. Corals are their favorite ornaments, and large quantities are imported every year.

VII.

CORAL REEFS.

1. Circular Islands.—The attention of seamen and navigators has long been attracted by the number of circular islands in the warm parts of the Pacific and Indian oceans. Generally each one of these circular islands contains a lake of quiet water extending almost to its outer shores, so that the island looks like a fairy ring of land floating in the ocean, and adorned with tropical trees and plants.

2. What are Coral Reefs?—Happily for the boys and girls of the present day, this subject, with other equally fascinating branches of science, has now been studied by naturalists, who give us the rich results of their labors. It seems scarcely possible that the dainty, beautiful corals which we have just examined can have anything to do with the making of islands, but nevertheless we find this to be the fact. Coral reefs are vast masses of coral which have grown in tropical oceans, where there is a strong current in the warm water. Their formation must have been slow, yet they sometimes extend hundreds of miles. Many parts of our solid continents are now thought to have been formed from coral reefs.

3. The Beginning of a Reef.—Let us now try to picture to ourselves the beginning of one of these reefs, and by following its growth step by step we may at last understand how it has been formed. There are hills and valleys on the bottom of the ocean as well as on the land. We will fancy that some young coral polyps which have been swimming about in the sea settle on the sides of one of these hills, and begin to grow and spread all around the hill. They will increase also by the deposit of eggs until they form a circular wall.

4. As the coral wall grows, the lower polyps and the inner ones die, their skeletons forming a solid foundation for all that grow above them. There may be only about an inch of living coral on the outside of the reef.

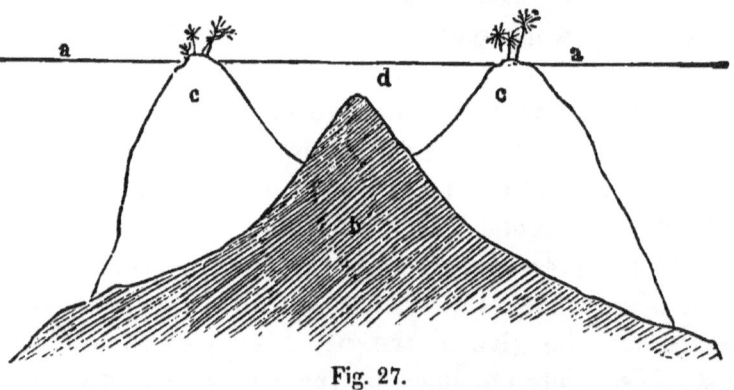

Fig. 27.

a, a, Surface of the Water; *b,* Natural Bed of the Ocean; *c, c,* Coral Formation; *d,* Lagoon.

5. How the Lagoon is Formed.—These walls rise nearly straight, and you will see that in doing so they enclose a circular basin of quiet water, and now you can understand why it is that a coral island mostly has a lake in the centre, as is shown in Fig. 27. The lakes are called lagoons.

6. Different Varieties of Coral found at Different Depths. —The bottom of the wall is formed of brain-coral and other solid kinds which live only in deep water, and these die when a certain height is reached. The formation of

the new island does not stop with their death, however. The wall having now reached the proper height to suit branching corals, which require shallower water, their young polyps will settle upon it, and finish the structure. We might suppose a reef formed of branching corals would be open and unsubstantial, but in their growth the branches are thickly interlaced. The spaces between them become filled with substances floating in the ocean, and with pieces of coral which are broken from the reef by the fierce dashing of the waves. These fragments of coral suffer no serious injury by breaking, but if lodged in some favorable spot they continue to grow, and together they form a solid mass, stronger, perhaps, than any stone masonry.

7. **The Sea not too Rough for the Polyps.**—The outer edge of the wall is steep and abrupt. Soundings taken just outside show very deep water. In this portion of the wall the corals live and thrive, always supplied with clear water, and an abundance of food brought by the rapid current. The breakers dash against it with such fury that apparently the hardest rock must in time yield to the tremendous force of the waves. But, strange as it may appear, the soft jelly-like bodies of the polyps give to the reef the power of resisting the billows.

8. **The Inner Surface of the Wall.**—The inner surface of the wall slopes gently to the land, and being washed by quiet waters often containing sand and mud, it is not favorable to the growth of polyps. Still, there are certain kinds of coral which thrive within the lagoons, and some of these are exceedingly brilliant and beautiful.

9. **How the Island is Raised above the Sea.**—The coral polyps die before they reach the surface of the ocean, as no corals can live out of water. The remainder of the

island is built up by shells, pieces of broken coral, sea-weed, and other floating materials which are washed upon it, and raise the wall higher and higher. The never-ceasing action of the waves grinds up these shells and broken coral, until at last they form a soil of sand and mud which is now ready to receive any seeds that may float on the water or be brought by the winds and the birds. The seeds take root in the new soil, and young plants begin to appear on the glistening white surface.

10. **The Vegetation.**—Cocoa-nut-trees are often the first to appear among these plants, the large nuts floating upon the ocean having lodged on the shores and found the warmth and moisture well suited to their growth. Other kinds of palms and pineapples also grow on these reefs.

11. **The Soil.**—The soil is thin, seldom being more than six or eight inches deep, but as the top of the reef is somewhat open and honey-combed, the crevices become filled with the soil we have described, and they make good holding places for the roots of large trees. As the plants drop their leaves and decay, the soil is enriched little by little, and the island is fitted for the home of various animals and birds, which in some mysterious manner find their way to these lonely spots far out at sea. In time our coral reef may become a beautiful tropical island fringed with waving trees and plants, and inhabited by man.

12. **Atolls.**—These circular reefs are called "atolls," and they seldom form complete rings. There is generally an opening into the lake on the side most sheltered from the wind. A safe harbor in mid-ocean is thus made, in which vessels may take shelter, but it requires an expert navigator to pass the perils at its entrance. In Fig. 28 is a pretty little coral island with ships in its lagoon. If a

lake is entirely enclosed by the coral wall, it may in time be changed to fresh water by the rains that fall into it.

13. **Theory of Coral Reefs.**—Coral reefs often extend to a depth of many hundred feet below the surface of the ocean, and formerly persons were puzzled to know how they could have been formed in such deep water, as reef-building corals do not thrive at a greater depth than one hundred and twenty or one hundred and eighty feet.

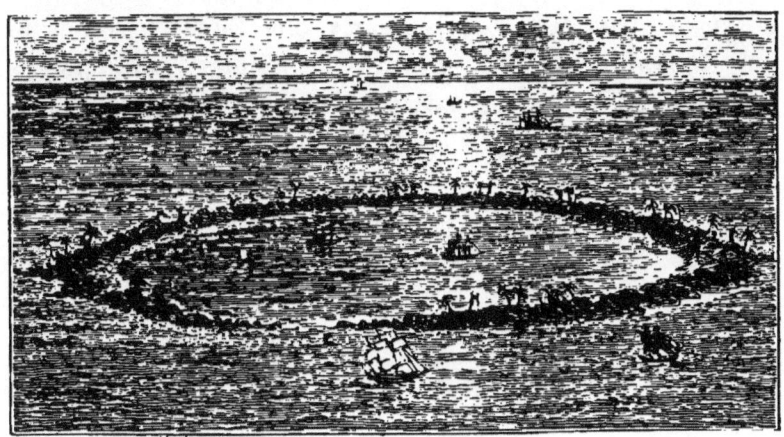

Fig. 28.—An Atoll.

This puzzling question was settled by the late Charles Darwin, who first showed that coral islands occur where there has been a gradual sinking of the bottom of the ocean. The theory is now generally adopted that as the growing reef rose in height, the foundation sank slowly, and in this way the upward growth was partly counteracted; consequently, the proper depth of water was secured, and the reef appeared to be stationary, whereas it was really growing upward.

14. When a coral reef rises above the surface of the

ocean, we may know that the coral, which grew under water, has been lifted above the level of the sea by a rising of the ocean bed since the reef was formed.

15. **Fringing Reefs.**—"Fringing reefs" are those which extend along the shores of continents and islands. There are usually openings or breaks in fringing reefs directly opposite the mouths of rivers and fresh-water streams, as the corals cannot endure currents carrying mud or sediment. Perhaps the grandest reef to be found in any part of the world is the one extending along the north-east coast of Australia. It is nearly one thousand miles in length, and proves to us that the helpless coral polyps have played no trifling part in the formation of our earth. All they have accomplished has been done merely by their living and growing.

VIII.

CTENOPHORA.

DAINTY MORSELS FOR THE WHALES.

SUB-KINGDOM, CŒLENTERATA : CLASS, ACTINOZOA.

1. **Ctenophora.**—Did you ever think how hard it would be to describe a soap-bubble to a person who had never seen one? It would even be difficult to paint a picture that would convey an idea of its delicate beauty. It will be quite as difficult to describe to you a class of animals almost as fairy-like as soap-bubbles, although they swim about in the ocean, and are honored with the high-sounding name of *ctenophora.*

2. **How shall we Pronounce the Word?**— At the first glance ctenophora may look like a hard word, but drop the "c," and you will find it quite easily pronounced— "te-noph'-o-ra." Were it possible for you once to see these charming creatures darting about in their native sea-water, their name henceforth would have a pleasant sound, and even a pleasant look, recalling to your minds lovely images of floating balloons and fairy bubbles.

3. **How shall we see Them?**—Ctenophora are too small and inconspicuous to be seen at the distance we usually are from the surface of the ocean, so the best way to observe them is in a large glass jar. On a calm day a jar of water dipped from the surface of the ocean may contain some of these beautiful creatures, although perhaps several jarfuls will have to be raised before the search is successful.

3

4. Transparency of the Ctenophora.—Upon looking close-ly at the little captives you will find them to be jelly-like, melon-shaped bodies, with bands running from end to end like the ribs on a melon. They are almost transparent, and if it were not for the prismatic colors that play upon

Fig. 29.—CTENOPHORA.

their sides as they glide through the water we could scarcely see them. If the ctenophora sporting about in the jar should swim in between you and any object be-yond the jar, you can see the object distinctly through their transparent bodies. Fig. 29 shows the form of one of the ctenophora, but it gives no idea of its delicacy.

5. **The Soft Bodies.**—The soft bodies of the ctenophora and their manner of life may remind you of jelly-fishes. Still, their structure is far more complicated, as we may observe through the clear substance of which the body is composed. When taken from the sea-water they lose their shape, and nothing is left but a film which is almost invisible.

6. **Jelly-like Animals could Live only in Water.**—The thought has perhaps already occurred to you that such animals as these, with jelly - like bodies, could live no-where but in the water. Many of them have no means of pursuing or of catching their prey, and they obtain only such food as is floated to them by currents in the water.

7. **Food.**—Although the ctenophora look so fairy-like, they devour a large number of animals, and they seem to prefer their own kindred. The mouth is at the upper end of the body, and when it is open, the food floats in and is quickly digested. In addition to the cavities nec-essary for digesting food, there is a set of canals within the body for the circulation of water.

8. **How Ctenophora Swim.**—The ctenophora swim about with exquisite grace, and yet they have no arms, no legs, no fins, to swim with. What need have they of any such organs? Their cilia are quite sufficient (the word cilia, you remember, means eyelashes). Those eight stripes we see running from one end to the other in Fig. 29 are bands of muscles on which are arranged comb-like fringes of cilia, which wave rapidly in the water, and give to the animal its lively motions. Indeed, it seems as if the fairy-like creature could not keep still. How can it keep still when these impatient cilia are striking the water? They send the little thing round and round, darting up and

down, till we wonder which way it will go next. The
cilia are worked by muscles under the control of the animal, and are to the ctenophora what oars are to a row-
boat.

9. **Study of Cilia.**—These eight bands of cilia add greatly
to the beauty of the dainty creatures. Their rapid motion
separates the rays of light that fall upon them, and produces down each band a flash of rainbow colors. In fact,
the cilia are so important and characteristic a feature of

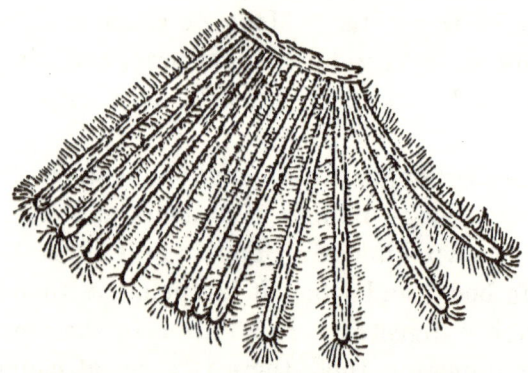

Fig. 30.—CILIA ON THE GILLS OF A MUSSEL.

the ctenophora that we should do well to become perfectly familiar with them. The appearance of these hair-
like organs is much the same wherever they are found,
and they show very distinctly on the gills of the mussel
(Fig. 30). These gills are fringed with countless cilia,
which under a microscope may be seen in rapid motion,
producing a continual current of water in one direction.
Their motion is regular, like that of the heart. The little
plates forming the gill lie side by side naturally, and
unless we looked very closely we might think the gills
consisted of only one piece. The plates are pulled apart
in the drawing to show the cilia more distinctly.

10. It is interesting to notice the various uses of cilia in the different positions in which they occur. Sometimes, as in the ctenophora, they propel the animal by striking the water like a multitude of tiny oars. Sometimes they surround the mouth, "and by their incessant action produce a small whirlpool, into which the food is sucked." In other cases their office seems to be to supply the needful air by keeping up a continual current of water, which contains as much air as these animals need. On the other hand, we must not imagine that cilia are confined to the lower animals living in the water. They serve important uses even in our own bodies. For instance, the air-passages leading to our lungs are lined with cilia, which are constantly lashing the air and beating back particles of dust and other impurities which it contains. Were it not for the cilia, these impurities would reach our lungs, and produce irritation there.

11. **Food for Whales.** — The beautiful ctenophora, idly sporting in the water, and seeming to have no aim but enjoyment, are far from useless, since they form the chief food of the Greenland whale. Do you not think these are dainty morsels for whales to feed upon? There must, however, be a good deal of nourishment in their transparent bodies, for the whales grow enormously large and fat. No doubt it takes a great many ctenophora to make a meal for the monsters. Large shoals of them are met with in arctic seas, and as the whales swim through the water with their great mouths hanging open, they catch the ctenophora on their whalebone fringes, and swallow a mouthful at a time.

12. **Their Abundance in the Ocean.** — In certain parts of the Arctic Ocean the water is of a grass-green hue, and is quite opaque. It is commonly spoken of as the "green

water," and its peculiar appearance is caused by the immense number of ctenophora it contains. These frolicsome little beings, living so thick and close as to color the water, are too small to be seen without a microscope. The

Fig. 31.—CTENOPHORA AND PHOSPHORESCENT FISHES.

rose-colored idyia, another species of ctenophora, is three or four inches long. It sometimes occurs in such numbers as to tinge large patches of the sea with its rosy color.

13. **Phosphorescence.**—All the ctenophora are phosphorescent. They are abundant on our own coasts, and are

often left on the shore at low tide, yet their beauty can only be seen as they glide daintily through the water. The eggs of some species escape singly, others are laid in strings or masses of jelly, and the young ones hatch out in the same form as their parents.

14. **Beauty of the Ctenophora.**—A jarful of sea-water dipped from the end of a pier one bright summer day contained four ctenophora, and made a whole party glad for an hour. It was a great delight to watch these little creatures darting hither and thither, sinking and rising again, or resting on their oars, according to their own sweet will. Sometimes we could not see them at all, though we knew they were in the clear water before us; then a flash of bright color appeared, and we followed their devious course by their glitter and sparkle.

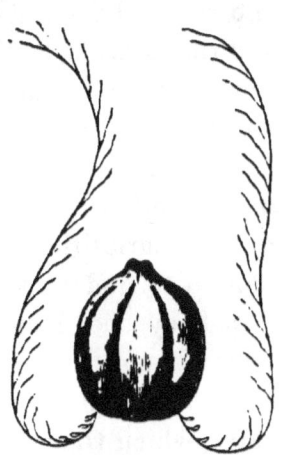

Fig. 32.—PLEUROBRACHIA.

15. **Pleurobrachia.**—One of these captive ctenophora was smaller than the others, and more nearly spherical. It belonged to the species *Pleurobrachia*, which you will see represented in Fig. 32. This was our especial favorite. At times it would throw out two long, slender tentacles, which were ornamented on one side with delicate tendrils. Upon some sudden fancy of the animal these tentacles were instantly drawn in out of sight, while at the next moment they were floating behind it for nearly half a yard. One might have supposed the exquisite creature was amusing itself by trying in how many different ways its tentacles could be curved and twisted.

IX.

STAR-FISHES.

SUB-KINGDOM, ECHINODERMATA : CLASS, ASTEROIDEA.

1. **Favorite Haunts of the Star-fishes.**—Those of you who go to the sea-shore in summer have perhaps discovered that star-fishes like rocky coasts the best. They are found most abundantly where the crevices between the stones afford good hiding-places for themselves and for the animals upon which they feed. They do not thrive upon muddy or sandy bottoms, and boys and girls hunting for curiosities upon such beaches are often disappointed to find no star-fishes.

2. They spend most of their time creeping over the rocks, though they love to be where the tide will ripple over their bodies and keep them well supplied with sea-water, which they depend upon for their oxygen. Those poor, half-dead star-fishes which we sometimes see in a pitiful condition on the beach will often revive if placed in sea-water, or, if left on the beach, the next high wave may restore them by carrying them out to sea again.

3. **How shall we Preserve them?**—Our dried specimens are yellow, but when alive, star-fishes are of a dull-red color, sometimes tinged with purple. They seem plump and fat on being taken from the ocean, but they are only puffed up with water, and if you watch them closely you will see the water oozing out all over the back. No doubt

you have learned how tedious and discouraging it is to attempt to dry star-fishes. You have perhaps been obliged to go home, as many before you have done, and leave them still drying in the sun. It may help you to know that a very good way is to dip them once or twice in boiling water before putting them in the sun or in a warm oven to dry.

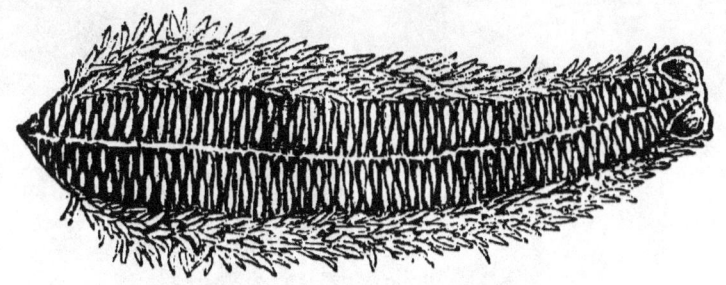

Fig. 33.—Under Side of Ray, showing the Hollow Tubes and the Limestone Plates of the Skeleton.

4. Broken Arms replaced. — Our common star-fish has five hollow rays or arms, extending from the centre like a star. If any of these rays are broken off, others grow in their places. It is a singular fact that these animals can break themselves to pieces, or throw off their rays, when they become alarmed.

- 5. The Skeleton.—Star-fishes glide along smoothly, and without apparent effort. They bend their bodies into various shapes to fit the inequalities of the surface over which they creep, and in order to do this they require a movable skeleton. See how beautifully Nature has provided for this necessity by forming the skeleton of thin limestone plates, so joined as to admit of slight motion. These plates are represented in Fig. 33, which is the under side of a ray, and the end having been broken off, we can see the two hollow tubes which the ray contains.

3*

6. The Upper Surface—The Madreporic Body.—Look now at the upper side of your star-fish, and notice the knobs and short spines with which it is covered.   If the ani-

Fig. 34.—Star-fishes (Lower One showing Under Side and Tube-feet).

mal is alive we may see between these spines tiny forks, with two prongs that are constantly snapping.   The use of the forks is not perfectly understood; they sometimes catch small prey, and they may also be useful in removing particles of matter that would otherwise choke up the pores on the surface. The first thing your bright eyes will discover is probably the round spot near the middle of the back and between two of the rays. That is called the "madreporic body," and it is an interesting object. Examine it with your microscopes, and try to think what those tiny holes can be intended for. It must

Fig. 35.—DINING UPON AN OYSTER.

be a sieve. Yes, it is a sieve, admitting water into tubes which run to the end of each ray. During life the madreporic body is bright-colored, and it strains all the water that enters the tubes, so there is no danger of their becoming choked.

7. **Singular Manner of Feeding.**—Now if we turn our star-fish over we shall find its mouth on the under side. This is an important organ, too, for star-fishes busy themselves continually with eating. They are especially fond

of live oysters and clams, and they have the oddest way
of eating them. They turn their stomachs right out into
the oyster shell, surrounding the soft body of the oyster,
and sucking it up. When the star-fish feeds it not only
bends its rays into a cup shape to hold on to its prey,
like the one in the picture dining upon an oyster, but
multitudes of tiny suckers spring up to help, and the prey
finds escape impossible. Oysters generally close their
shells so quickly in time of danger that we cannot understand why they should allow the sluggish star-fishes to
catch them napping. It has been suggested that the star-
fish drops into the shell some liquid which paralyzes the
oyster, but this no one knows. So you see the star-fish,
without any tools, is able to help itself to raw oysters.

8. **Tube-feet.**—The way in which star-fishes walk is also
curious. It will repay you well to examine the next living
star-fish you find, and notice the odd manner in which it
glides along. On the under surface of each ray is a
double row of hollow tubes, which squirm and grope
around like a multitude of worms. As these are the organs by which star-fishes move, they are called tube-feet.
They are lengthened and enlarged, much as the tentacles
of sea-anemones are, by filling them with water. For this
purpose each tube-foot is connected with a little round
bag filled with water from the water-tube running down
the ray. When the bag contracts it forces water into the
foot, which reaches forward and attaches itself by a round
sucker on the end to the surface over which the animal
wishes to move. In this way one sucker after another is
stretched out to cling to the surface, and as the suckers
are shortened again by expelling the water, the body is
dragged forward. Fig. 36 shows the interior of one of
the rays. The tube-feet, *g*, are shrunken up quite short,

which makes the water-bags, *h*, all the larger. Notice the mouth, *a*, the stomach, *b*, and the intestine, *c*.

Fig. 36.—INTERIOR OF RAY.

a, mouth; *b*, stomach; *c*, intestine; *d*, upper surface; *e*, limestone plates; *f*, ovary; *g*, tube-feet; *h*, water-bags.

9. The double rows of tube-feet are set in a deep groove. In your dried specimens the tube-feet have shrivelled up and fallen away, and in the grooves you will probably see a number of delicate plates arranged side by side in two rows. These are called "ambulacral plates," and they are sufficiently far apart to allow water to flow out between them from the water-bags into the tube-feet. Notice this in Fig. 33. On the outer edge of the rays is a number of stiff spines.

10. **Other Organs.**—Star-fishes have a liver and intestines. Their organs do not lie wholly in the central portion, but extend into the five hollow arms. They also have nerves, which surround the mouth and pass down each arm, where they end in a red eye-speck. This arrangement gives to star-fishes five eyes. They are not perfect eyes, however, and it is probable that they can see but little. Star-fishes are said to be careful of their eggs, carrying them with the suckers near the mouth.

11. **Destruction of Oysters.**—The star-fish's fondness for fresh oysters is a serious matter to the oyster-grower, and

causes him to lose large quantities of his valuable prop-
erty. It is estimated that the damage every year to the
oyster-beds between Staten Island and Cape Cod amounts
to $100,000. Large numbers of star-fishes sometimes ap-
pear suddenly and unexpectedly upon the shores. They
seem to be washed in from the deep sea, and, settling
upon the oysters, they begin their work of destruction,
and consume many bushels in a short time. These attacks
occur chiefly in the latter part of summer or early in the
fall, and are much dreaded by the owners of oyster-beds.

12. **Some Uses of the Star-fish.** — The oysterman has
learned the value of these destructive pests for manure,
and those dredged from oyster-beds are now saved for
fertilizing purposes. We might also attribute to the star-
fishes a certain usefulness as scavengers of the ocean,
since they eat all sorts of animal substances, dead as well
as living, and do their full share towards keeping the
waters pure.

13. Some kinds of star-fishes have long feathery arms,
and are much more beautiful than our common ones
which we have been studying.

X.

SEA-URCHINS.

SUB-KINGDOM, ECHINODERMATA : CLASS, ECHINOIDEA.

1. Sea-urchins.—What funny, prickly creatures the sea-urchins are ! A person might easily mistake them for green chestnut-burs scattered on the beach, and, glancing up hastily, he might almost expect to find the overhanging

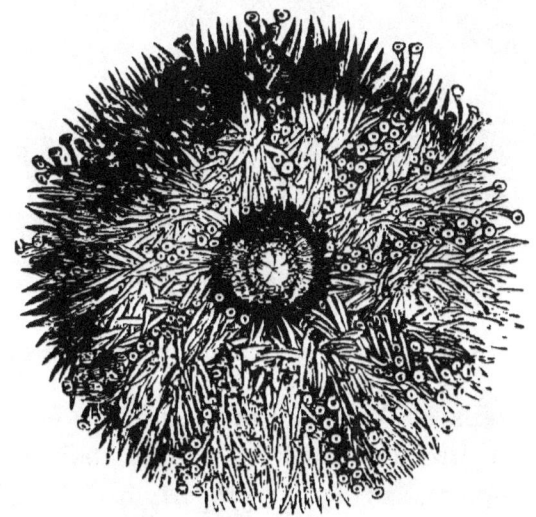

Fig. 37.—UNDER SURFACE OF A SEA-URCHIN, SHOWING ROWS OF SUCKERS AMONG THE SPINES.

branches of a great chestnut-tree. By this time, however, the prickly green things may have stretched out their purple suckers and begun to drag themselves over the beach. This motion prompts us to place them among the animals.

2. **How are they like Star-fishes?**—We have seen the same method of travelling practised by our old friends the star-fishes, yet surely these round creatures can be nothing like star-fishes. But that is just what they are like, and I think we shall soon discover a close relationship between the two. We might almost say that the sea-urchin (Fig. 37) is a star-fish that has got up in the world, and, folding its rays together side by side, has concluded to live henceforth shut up in its beautiful round box.

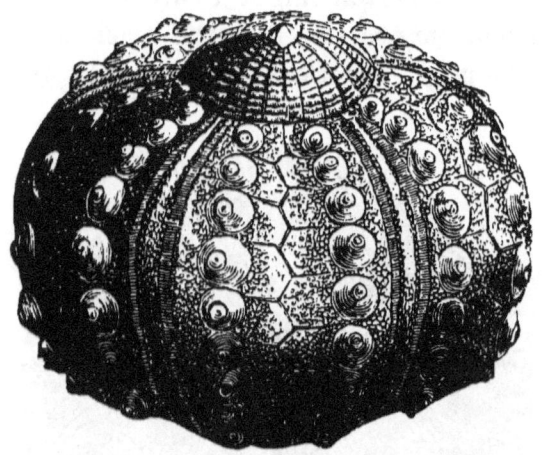

Fig. 38.—Shell of a Sea-urchin without Spines.

3. **The Shells.**—We sometimes find the empty white shells of sea-urchins which have lost their coating of prickles, or spines, as they are properly called. The shells are very elegant, being scarcely thicker than an egg-shell, and ornamented with rows of dots and knobs with open lace-work between. This shell is not one globular piece, as you might suppose, but it is made up of several hundred little plates exquisitely fitted together, and forming a true mosaic, as seen in Fig. 38. On the inside of the shell you can easily see the lines where these plates are

joined, and you will surely be charmed with the double
rows of lace-work which divide the shell into five equal
sections. Let us see what they mean.

4. **Plan of the Sea-urchin.**—If we should place a star-
fish on the table with its mouth down, and bend its rays
backward until they meet together on top, and the edges
of the rays touch each other, we should have the general plan of a sea-urchin. Do not imagine that star-fishes
ever do turn into sea-urchins. This is merely intended to
show you the similarity of their structure.

5. According to this arrangement the double rows of
perforated plates would represent the middle of the rays
of the star-fish where
the tube-feet are situ-
ated, while the broad
belt of knobs corre-
sponds to the strip of
spines on both edges
of the ray.   The
mouth would be un-
derneath, and you
would of course look
for the five eye-specks
on the top where the
ends of the rays meet.
You will also find the
madreporic body at

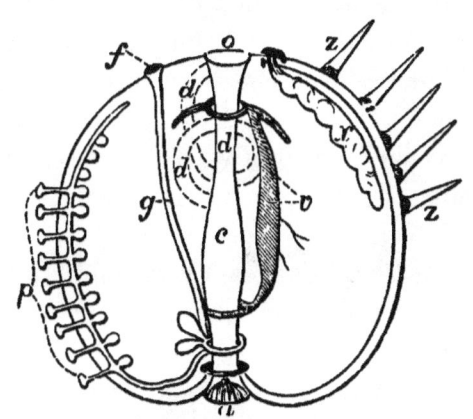

Fig. 39.—Section of a Sea-urchin.

a, mouth; c, stomach; d, intestine; o, anus;
v, heart; f, madreporic body; g, main water-
tube; p, tube-feet; z, spines.

the top, a little on one side. This small sieve, which is
so interesting in the star-fishes, performs the same service
for the urchin, and prevents the entrance of any sand or
other solid substance into the five tubes that pass under
the holes in the shell. Through these holes are pro-
truded the double rows of tube-feet, just like those we

have studied in the star-fish, and they are worked in the same manner. It adds much to the beauty and interest of the shell to know that these lovely fine dots are openings through which the tube-feet are supplied with water from inside the shell. In the diagram (Fig. 39) the madreporic body is shown at f, and the tube, g, carries water to supply the tube-feet and their little water-bags, which are shown at p. Sea-urchins move by means of their tube-feet, which may be lengthened so as to extend far beyond the spines.

6. **Growth of the Shell.**—The shell fits the animal exactly in its infancy, and must still serve it in old age, for urchins never cast off their coats as crabs and lobsters do. Being formed of many small pieces, the shell grows a little all over. Each plate is surrounded by living flesh. This flesh secretes lime from the sea-water and deposits it round the edge of the plates, thus increasing the size of the shell uniformly. After sea-urchins die the spines drop off; the shell is so frail that it too is soon broken, the plates falling apart.

7. **The Spines.**—Do not omit to look at the spines with your microscopes, and see what beautifully carved columns they are (Fig. 40). Falling about on the back of the urchin, they remind one of a sadly neglected graveyard, with its tottering monuments. Each spine is hollowed at the end to fit a knob on the shell. This forms a perfect ball - and - socket joint, which is supplied with delicate muscles to move the spines. As the creature travels along, the spines are constantly in motion, and they look as if they too wanted to help. In some species the spines are very large, and are used for slate-pencils. We should think it quite luxurious to have such artistic pencils, yet many boys and girls in out-of-the-way places, especially where fishing - vessels bring home curiosities

from foreign shores, have puzzled over their examples and written their copies with these elegantly fluted spines.

8. The Teeth.—Did you notice the white spot in the middle of Fig. 37, also the pointed beak near the top of Fig. 38 ? Both of these spots show the five white teeth which come together in a point, and may be extended beyond the shell just as they are in Fig. 38. You will observe what a great step forward the sea-urchin has made. We have found nothing like teeth before in the lowly creatures we have been studying, and here is the urchin, armed with five hard white teeth, having sharp cutting edges like a rat's teeth.

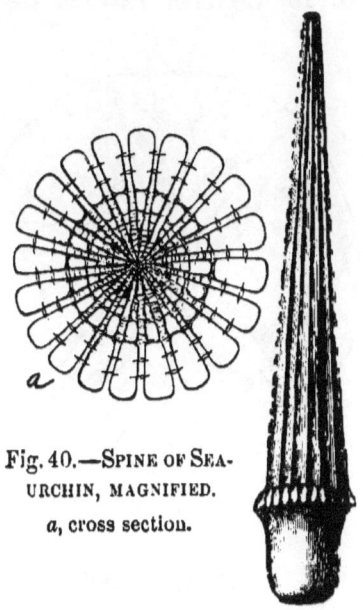

Fig. 40.—SPINE OF SEA-URCHIN, MAGNIFIED.

a, cross section.

Each tooth has a separate jaw of its own, and is worked by its own muscles. This singular arrangement has attracted much attention, and from the shape of the jaws and teeth it is known as "Aristotle's lantern."

9. Internal Organs.—The sea-urchin is well supplied with organs (as we may see in Fig. 39)—the mouth, *a,* the stomach, *c,* the coiled intestine, *d,* and the anal opening, *o*—whereas our studies heretofore have been about animals with a simple sac for a stomach, and all the refuse of their food was returned through the mouth. This highly favored individual has also a heart, *v,* and blood-vessels, although the blood which passes through them is quite different from that of higher animals.

10. Sea-urchins as an Article of Food.—The sea-urchins of the Mediterranean are larger than ours, and are used for food, either raw as we eat oysters, or boiled. They

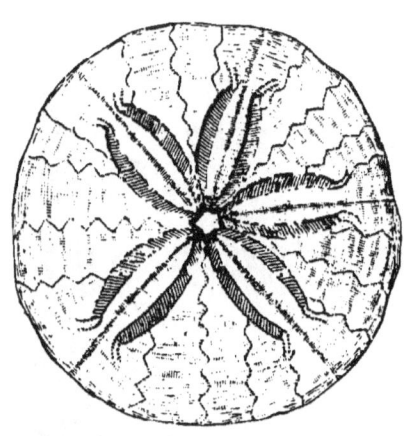

were a favorite dish with the ancient Greeks and Romans. Bunches of their eggs are also offered for sale as food in the Italian cities. The eggs pass out of the shell through small openings near the madreporic body, and they are often seen on top of the shell, surrounded by spines which have been drawn together to hold them.

Fig. 41.—SAND-DOLLAR.

11. Echinoderm Defined.—Young people like to use the proper names for things, and now that we know all this about the sea-urchin we will give it its right name, the *echinus*. In your reading you will also meet with the word *echinoderm*, and it will give you pleasure to recognize it as an old acquaintance. Echinoderm means spiny-skinned. It is the general name given to star-fishes, sea-urchins, and their relations, most of which have prickly coverings.

12. Boring in Rocks.—The echinus has a curious habit of boring holes in hard rocks. It sinks in the hole for a considerable distance, and looks well satisfied with its snug retreat. It is not understood how the rock becomes worn away, unless it is by a rotary movement of the body. Constant dropping, we know, wears a stone, and constant turning and twisting may do the same. There is no doubt the hole is made by the animal which occupies it, as it fits exactly, whether the occupant be large or small.

13. It is amusing to watch the echinus in shallow water drag itself along by its tube-feet, and sometimes hide itself by drawing together pieces of sea-weed and gravel.

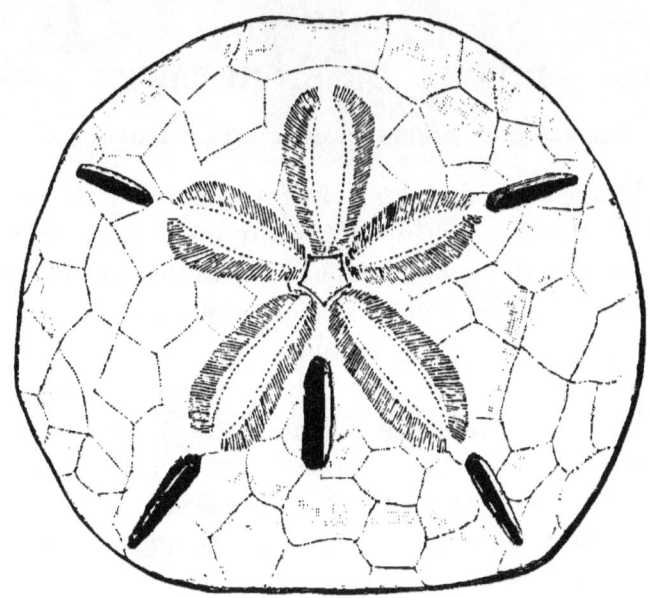

Fig. 42.—KEYHOLE-URCHIN.

14. **Varieties of Sea-urchins.**—In visiting a good museum you will be surprised to see how many different varieties of these creatures there are. Some species are flattened, and pass by the name of *sand-dollars* (Fig. 41), *keyhole-urchins* (Fig. 42), etc. During life the skeletons of these animals were covered with skin, and furnished with a furry coat of little spines and small tube-feet.

XI.

CRINOIDS, OR STONE-LILIES.

SUB-KINGDOM, ECHINODERMATA : CLASS, CRINOIDEA.

1. **Where Crinoids grow.**—While examining the sea-urchins at the museum do not fail to hunt up the crinoids
also. We shall have to be content with this way of studying crinoids, as the living ones grow on rocky beds in the
deep ocean. They are obtained only by dredging, and
few of us will ever have an opportunity to see them.

2. **Why they are called Stone-lilies.**—Crinoids are attached during the whole or a part of their lives to the sea-
bottom by means of a jointed stalk which is so flexible as
to bend freely in any direction. At the upper end of the
stalk is the cup-shaped body, with its waving arms, which
may be folded together like a flower-bud, or spread open
like the petals of a full-blown lily. Swaying to and fro
in the bright water, this curious animal closely resembles
a flower tossed by a gentle breeze, and as it really has a
hard skeleton throughout, " stone-lily " is not a bad name
for it.

3. **Crinoids compared to Star-fishes.**—Let us imagine a
star-fish supported in this way upon the end of a long
stalk, and we shall have a pretty good idea of a crinoid.
In comparing the two we must invert the star-fish, however, as the mouth of a crinoid is on the upper surface,
whereas in the other echinoderms the mouth is underneath.

The tube-feet, like-
wise, are on the up-
per surface of the
arms, but they are
not used for travel-
ling about as the
tube-feet of other
echinoderms are.
The grooves contain-
ing them are covered
with cilia which cre-
ate currents of water
towards the mouth,
and carry to it the
minute plants and
animals upon which
the crinoid feeds.

4. Skeleton of Cir-
cular Plates. — Like
star-fishes and sea-
urchins, these cousins
of theirs secrete lime
to form a solid frame-
work for their bod-
ies. The lime is de-
posited in circular
plates, which are sur-
rounded and held to-
gether by living flesh,
so that they bend ea-
sily. You can detect
these circular plates
in any part of the ac-

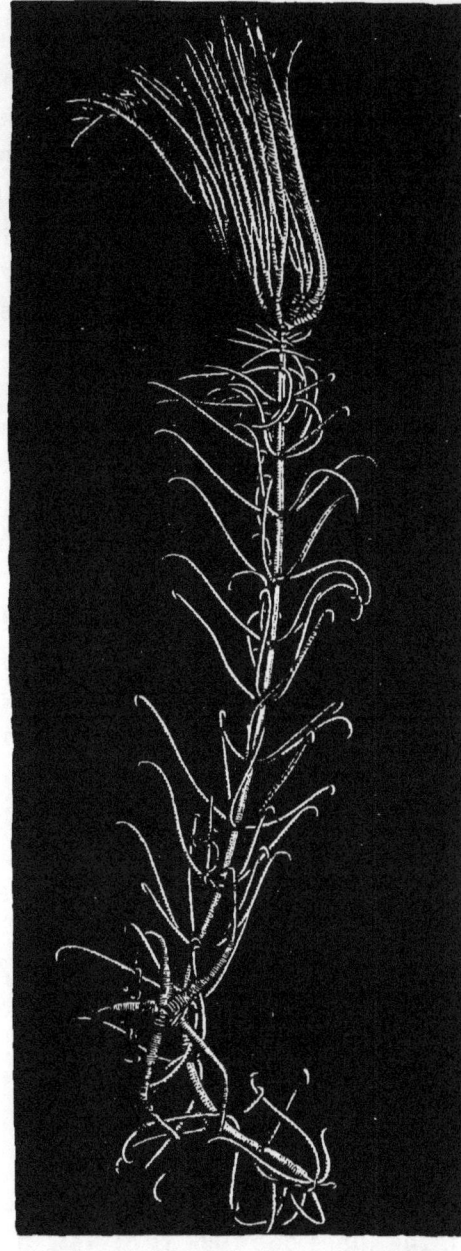

Fig. 43.—A Living Crinoid. West Indies.

companying picture.   Indeed, crinoids may be known by
the little rings of which they are composed.

5. **Free Swimming Crinoids.**—In some crinoids, as the
Comatula, or feather-star, the animal is fastened to the
ground only when young.   Later in life it drops from the
stalk, and is free after this to travel about.   It can swim
through the water; still, it prefers to remain quietly settled on some stone or sea-weed, waving its feathery,
bright-red arms while it feeds upon the little animals
floating around.   It now resembles a star-fish more than
ever, though it moves only by means of its flexible arms.

6. **An Ancient Family.**—The family of crinoids is very
ancient, and was perhaps at one time the most numerous
family which inhabited the sea.   Like some other old
families, it has almost died out.   There are but few species now living, and two or three of these have been only
recently discovered by scientific explorers while dredging
the deep waters of the Caribbean Sea and the Atlantic
Ocean.

7. **Fossil Crinoids.**—Fossil remains of crinoids are abundant in rocks, showing that in past ages they must have
lived in great numbers.   In France large beds of rock are
formed of their remains, and the same is true of many
other parts of Europe and North America.   The circular
plates of the crinoids were so loosely held together by
flesh that when the animal died they fell apart, and
these little disks which are now found in the rocks look
like button-moulds ornamented with beautiful patterns
and markings.

8. See how the crinoid stems are piled upon each other
in the limestone rock (Fig. 44), and notice the little hole
in the middle of each.   Their arrangement in the rock
(Fig. 45) is much more orderly.   No wonder that such

fine old crinoids as this should have been mistaken for petrified flowers.

9. **How Fossils came to be in the Rocks.**—Perhaps you will wonder how animals can be embedded in hard rocks.

Fig. 44.—CRINOIDAL LIMESTONE.

To understand this we must remember that many of our rocks are formed of sand or mud, which has become hard from the constant pressure of other layers of sand and mud that have accumulated above them. Most of these layers were formed underneath the sea. The rocks must have been in this soft condition when the animals died and were buried in them. As the rocks hardened, the solid parts of the animals were preserved in a stony bed, the hard rock fitting closely into every crack and cranny. When these rocks are split open we sometimes find the remains of the animal on one side of the crack, and a perfect impression on the other. These petrified remains are called *fossils*, and they tell us a fascinating story of the curious animals and plants that lived long ago.

4

10. **Records of our Earth's History.**—The true nature of fossils, and the causes which placed them in solid rocks, interested the poets and philosophers long before the Christian era. It is only within the last century, however, that they have been accepted as records of the history of our earth. There are many animals now entirely extinct of whose existence we should know nothing but

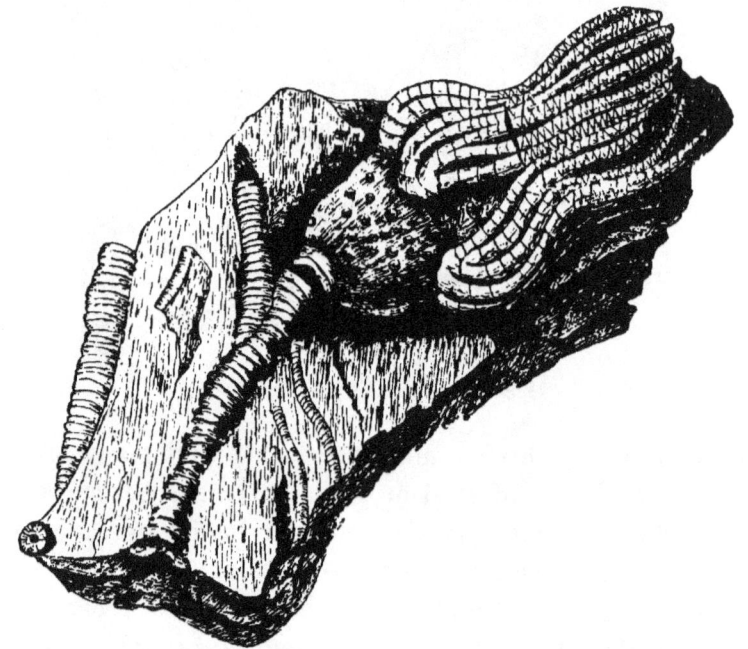

Fig. 45.—A Fossil Crinoid.

for their fossil remains. These relics of the past tell also of great changes from heat to cold in certain parts of the earth. For instance, the bones and teeth of elephants, rhinoceroses, and other animals that require warm climates are found in Siberia and in other cold countries, which shows that the polar regions were once much warmer than they are now. Again, on the other hand, remains of reindeer are found in Southern Europe, indicating extreme cold at another period in the earth's history.

11. So you see these fossils have wonderful secrets to tell. Strange, old-fashioned secrets, for the formation of the rocks has been very slow, and the animals buried in them must have died thousands of years ago. Crinoids and corals and shells which live only in the ocean are found in a fossil condition in the interior of the dry land, proving beyond a doubt that these parts of our earth must at one time have been beneath the sea.

12. Is it not a lovely thought that these delicate crinoids which beautified the ocean long before we were here to admire them are not utterly destroyed, but that some of their skeletons have been preserved and are waiting for us in the gray old rocks, if only our tastes are simple and cultivated enough to find them out?

XII.

SEA-CUCUMBERS.

SUB-KINGDOM, ECHINODERMATA : CLASS, HOLOTHUROIDEA.

1. **The New England Coast favorable for Growth of Sea-animals.**—The shore of Maine, you will remember, is very uneven, being broken by a succession of sharp promontories and quiet bays, and skirted with a fringe of lovely islands. Here is an endless variety of bold rocky cliffs, of secluded caves and quiet little pools, with the pleasing surprise of occasional short sandy beaches. We can scarcely imagine a shore better fitted than this to suit the various tastes of the sea-creatures, and our search here is pretty sure to be rewarded by finding sea-anemones, star-fishes, sea-urchins, sea-cucumbers, etc., besides a variety of shell-fishes. Sea-weeds also grow in abundance, coloring the water with their beautiful tints.

2. This is true of the New England coast as far south as Cape Cod, while below that point the sandy beaches of the Atlantic shore are not favorable for the growth of these animals. In addition to the loose sand which is washed up on the beach, the great number of rivers emptying fresh-water into the sea renders it still more unfavorable for their abode.

3. **Sea-cucumber.**—As found on the beach, a sea-cucumber would remind you of a leather bag, somewhat worm-like in form, with no hard shell, and marked with rows of warts down the sides like a cucumber (Fig. 46). The skin

is tough, yet it may expand and contract in such a way as to give these animals the curious power of changing their shape.

Fig. 46.—Sea-cucumbers (Holothurians).

4. **Changes into Odd Shapes.**—Upon watching the movements of a sea-cucumber you will be amused at the odd shapes into which it changes. It sometimes lengthens out its body like a worm, then drawing itself in tightly around the mouth, it swells out the other end of the body like a jug. Suddenly, tiring of this freak, it can make an hour-glass by contracting its body, as if a string were tied around the middle of it, with bulges above and below.

5. **Feathery Tentacles around the Mouth.**—The tentacles of a sea-cucumber form a feathery fringe around the mouth. Their number is usually ten, and they have the same curious power of changing their shape that we have noticed in the body of the animal. The mouth may be distinctly seen in Fig. 47, which represents another species of sea-cucumber. It opens into a throat leading to the stomach. The long intestine passes to the other extremity of the body.

6. Sea-cucumber an Echinoderm.—From the general appearance of the sea-cucumber you will scarcely suspect that it is one of the echinoderms, but, after watching it creep over the rocks, you can see the relationship. The tube-feet steal out noiselessly from the wart-like spots, as seen in Fig. 48, and the sea-cucumber travels just like a sea-

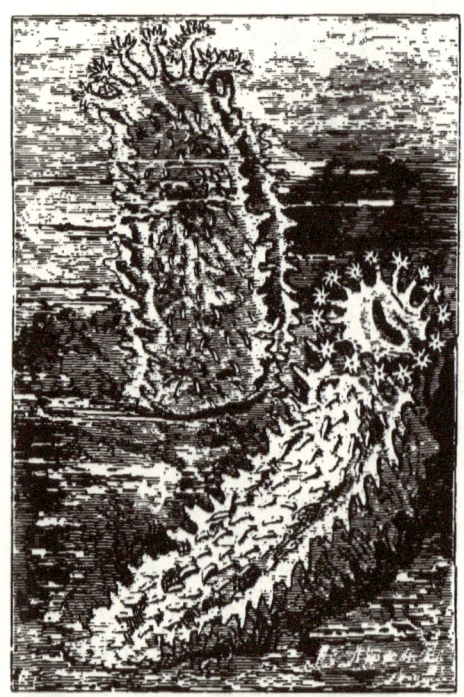

Fig. 47.—Sea-cucumbers.

urchin. The tube-feet are arranged on five muscular bands running from end to end, and dividing the body into five segments. The spaces between the tube-feet correspond to the spaces which are covered with spines in the sea-urchin. One species of sea-cucumber has the tube-feet all collected on the under side of the body (Fig. 49). It is called a "*sea-orange*," probably from the rough rounded markings on its surface. In those species which have no tube-feet the animal drags itself along by the aid of anchor-shaped spicules scattered through the skin.

7. Resemblance to other Radiates.—The madreporic body is not on the outer surface, as it is in other members of the family. It opens upon a little canal in the interior, which supplies the tube-feet with water. Although hidden from our view, this tiny sieve filters the water perfectly, and allows no irritating particles to enter the tube. The only resemblance to the other Radiates which we detect in these animals is in the arrangement of their tentacles, their tube-feet, and muscular bands.

8. Castaway Organs Replaced. —The sea-cucumber does not break itself to pieces as the star-fish does, but it has a peculiarity quite as remarkable: when alarmed it throws away various organs from the interior of the body, and, strange to say, these castaway organs are soon replaced by others.

9. Holothurians— where found. — Sea-cucumbers, or *holothurians*, as they are prop-

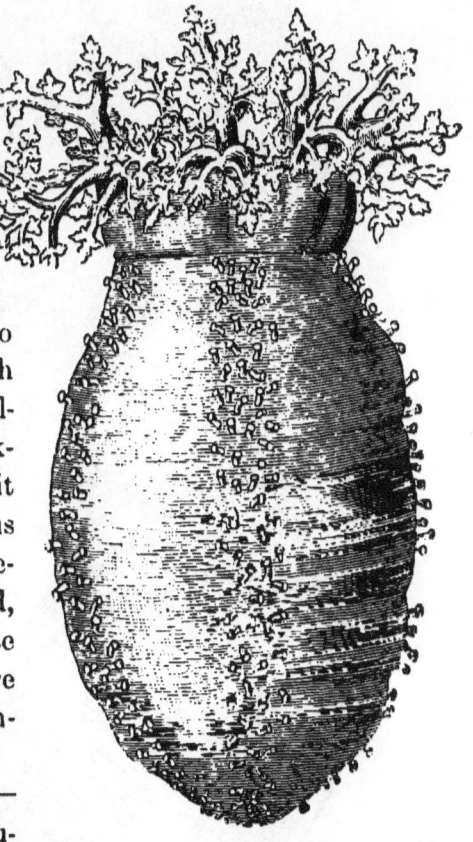

Fig. 48.—A SEA-CUCUMBER (PENTACTA FRONDOSA).

erly called, are most abundant in tropical seas, where they lie in the mud, or in shallow water, with their tentacles floating in expectation of prey. These creatures, as found on our shores, with their tentacles snugly stowed away, have no pretensions to beauty. One species, however, from the Pacific Ocean is described as being much hand-somer than the rest of its kind. The body is as transparent as glass, and of a lovely rose-color, with fine white

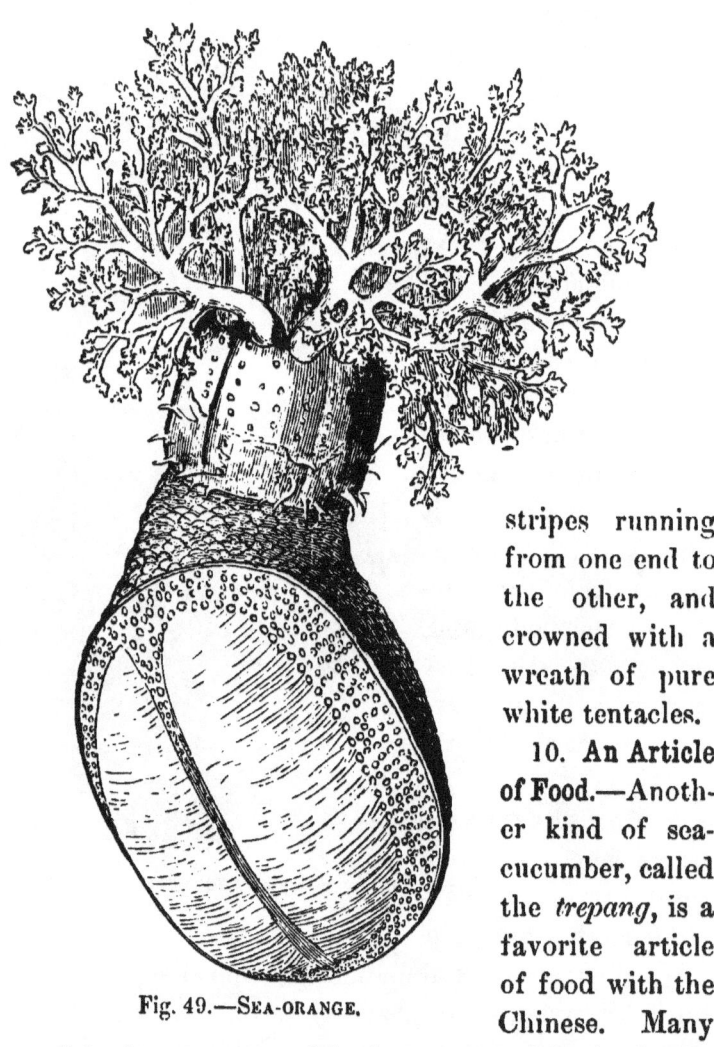

stripes running from one end to the other, and crowned with a wreath of pure white tentacles.

10. **An Article of Food.**—Another kind of sea-cucumber, called the *trepang*, is a favorite article of food with the Chinese. Many

Fig. 49.—Sea-orange.

thousand junks are engaged in the trepang fisheries in the Indian Ocean. The trepangs are caught with a harpoon as they creep over the rocks and corals, or, when the water is shallow, they are brought up by divers. While yet alive the animals are thrown into boiling sea-water, and stirred with a long stick. After being boiled and flattened with stones, the Malay fishermen spread them on bamboo mats, where they are dried and smoked; then they are packed and shipped to the Chinese market.

11. **Jelly Lumps in the Sea are often undeveloped Young.**— In strolling on the sea-shore we often find little lumps of clear, transparent jelly left there by the retreating tide. Many of these jelly lumps are the undeveloped young of the class of animals we have been studying; and if some time you should place a number of them in sea-water, and change the water frequently, you may have the pleasure of watching their development, and see what special forms they assume. These animals produce young ones in great abundance. It is necessary they should do so, or the race would soon die out, as they are devoured in such numbers by fishes that only a small proportion of them live to maturity.

12. **Animals Preying upon Others.**—The sea contains myriads of animals that prey upon each other, the larger ones eating the smaller; and we can form but little idea of the amount of life continually sacrificed for the support of that which remains. It seems almost marvellous that any of the delicate little ones should escape the hungry hordes that pursue them in this eager struggle for life.

4*

XIII.

EARTH-WORMS.

SUB-KINGDOM, ANARTHROPODA; CLASS, ANNELIDA.

1. **The Work of Earth-worms.**—Who would have thought the little earth-worm had any work to do in the world, or was of other use than to bait fish-hooks? Yet it has an important part to perform, and we are now told that the present fertile condition of the earth is largely due to the action of earth-worms.

2. If this is the case, we must look at these industrious workers more carefully. Having selected a fine large specimen, let us put it on a piece of white paper, where it will show to advantage.

3. **A Land Animal.**—This is the first animal we have examined which lives upon land. The simplest forms of life are always found in water, but from this point in our studies we shall sometimes take our specimens from the land, and the boys and girls all over the country will have an equal chance to obtain them. Even those who live in large cities can procure earth-worms.

4. **Study of the Illustration.**—Let us study for a moment the illustration of an earth-worm that we have here. The worm itself is shown at Fig. 50, *a ; b* is a small part of it magnified so as to show the bristles pointing backward. The egg of the worm, *c*, is curiously constructed, having a valve at one end. In *d* we see the young worm, which has opened the valve and is coming out of the egg.

5. The Body made up of Segments — Articulates. — The body of the worm tapers towards each end, so that we can scarcely tell the head from the tail unless we watch a worm as it is creeping. Notice all those little rings across the body, and see how they slip in and out of each other as the worm moves. These rings can be drawn so close together that a large worm will sometimes make itself very short. Does this creature look like a Radiate? Certainly it does not, and we will now learn that all animals which have the body made up of rings or segments extending crosswise belong to a class called Articulates.

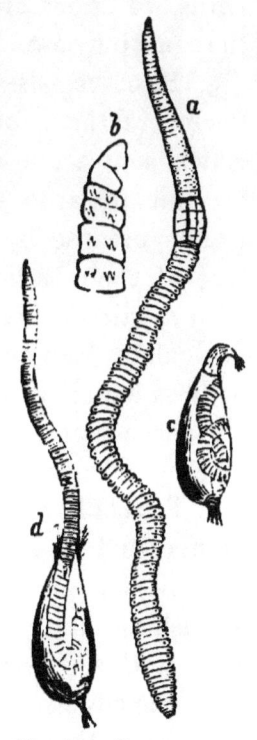

Fig. 50. — EARTH-WORM.

6. The Bristles. — The earth-worm contains from one hundred to two hundred of these rings, each of which is furnished with four pairs of bristles pointing backward. You can easily feel them with your fingers. The bristles assist in crawling, and prevent the worm from slipping back as the rings are contracted and expanded. Still, the worm can creep backward when it desires to, and you may have noticed how rapidly these timid animals draw back into their holes.

7. Organs. — Earth-worms have no distinct head or eyes. The mouth consists of two lips, and it has neither teeth nor tentacles. The semi-transparent body will enable you to see the food canal, extending from the mouth through the whole length of the worm, and enlarged in two places to form the crop and gizzard. Grains of sand and small stones are often found within the strong gizzard, where they probably act as millstones in helping to grind the food. Birds, we know, are in the habit of swallowing stones for the same purpose.

8. **Blood-vessels—Ganglia.**—We find no heart in these lowly creatures, but in its place a set of blood-vessels, which contract in such a way as to force the blood from the tail towards the head. It is supposed that earth-worms breathe by tubes opening upon the external surface of their bodies. Each one of the rings is supplied with a pair of nervous ganglia. This is true of all the Articulates. In animals of this group, each segment of the body is supplied with its own nervous ganglia. Ganglia are nerve-centres which consist of a mass of nerve-cells sending out nerve-fibres to other parts of the body.

9. **Food of Earth-worms.**—Worms live in burrows in the ground, and in making them they swallow an almost incredible amount of earth, out of which they take all the nourishing matter. They do not confine themselves, however, to this coarse diet, but feed upon leaves and stems, from the edges of which they suck off little bits, having first drawn them into their burrows for a distance of two or three inches. Leaves are also dragged in for plugging their burrows, and when they cannot get leaves for this purpose they sometimes pile up heaps of stone to close the entrance. This work is all done during the night.

10. **The Burrows.**—The burrows are often lined with a layer of fine earth, which seems not only to strengthen the walls, but to form a smooth surface for the worm's body. At the bottom of the burrow there is generally an enlarged chamber which contains small stones, and here the worms pass the winter rolled up two or three together in a ball.

11. **Castings made by Earth-worms.**—Now, if we want to know what becomes of the earth which is swallowed by worms, we have but to remember the rounded, worm-like heaps of earth called "castings" which are so thick among the grass, and on the untrodden parts of paths and drives, or in the flower-pots when a few worms have been dug up with our favorite house-plants. When a worm comes to the surface to empty its body it backs out of its hole, and ejects the earth which it has swallowed in spurts, first on one side, then on the other, until a little heap is formed, which hardens in drying. It is estimated that the quantity of fine earth thus carried to the surface in the course of a year would in many places form a layer one-fifth of an inch in thickness, amounting to a weight of more than ten tons on each acre.

12. **Vegetable Mould the Work of Earth-worms.**—Have you ever noticed the layers of different-colored earth that are exposed in digging a well or a cellar? The upper layer, you may remember, is mostly of a rich dark color. It consists of fine soil two or three inches deep, which has been sifted of stones and coarse materials, and is spoken of as "vegetable mould." This fertile layer is the work of earth-worms.

13. Charles Darwin estimated that the whole mass of vegetable mould which is spread over the surface of the earth passes through the bodies of worms once every four years, and in this way fresh masses of earth are constantly exposed to the influence of rain and wind. Worms also do much to enrich the soil by the great number of leaves and twigs drawn into their burrows.

14. "The bones of dead animals, the harder parts of insects, the shells of land mollusks, leaves, twigs, etc., are before long all buried beneath the castings of worms, and

are thus brought in a more or less decayed state within reach of the roots of plants."

15. The Earth Ploughed by Worms.—"The plough is one of the most ancient and most valuable of man's inventions; but long before he existed the land was, in fact, regularly ploughed, and still continues to be thus ploughed, by earth-worms. It may be doubted whether there are many other animals which have played so important a part in the history of the world as have these lowly organized creatures." The corals, indeed, have done more conspicuous work in constructing great reefs and islands, but these are mostly confined to the tropical zones.

16. How Stones and Pavements Disappear.—It is no new discovery that pebbles and cinders and even large stones lying on the ground in a few years disappear. Neglected and unused pavements also become covered with soil, much of which has been raised by these busy little workers. Such every-day wonders escape the attention of most of us, but Charles Darwin, while pursuing his studies and observations upon various subjects, still found time to notice the worms. He and his sons watched them for more than thirty-five years before he published the book which gives these interesting facts.

17. Destruction of Worms by Birds.—He says that worms often lie motionless for hours just beneath the mouth of their burrows, so that by looking closely their heads may be seen. If the earth or rubbish over the burrow be suddenly removed, the worm retreats rapidly. This habit of lying near the surface leads to great destruction. At certain seasons of the year the thrushes and blackbirds draw large numbers out of their holes. Watch a robin some morning hopping over the lawn, and see how it pecks, and pecks, at some object, finally bracing itself upon its tail, and pulling with all its might, as if determined to draw out its victim; but the worm holds on so tightly by its short bristles that it is no easy matter for the robin to capture it.

18. **Found all over the World.**—Earth-worms exist all over the world, in cold countries as well as in warm ones, and even in small islands far out in the ocean. They require some moisture, and during very dry weather, or when the ground is frozen, they retire to a considerable depth.

19. Large numbers of worms are often found dead on the pavements after a heavy rain. As earth-worms like moisture, it is scarcely probable these have been drowned. Darwin suggests that they were already sick, and that the flood may only have hastened their death.

XIV.

CRABS.

SUB-KINGDOM, ARTHROPODA : CLASS, CRUSTACEA.

1. **Crabs.**—Crabs are curious creatures. At the first glance we can scarcely tell which is the head. Notice the position of the eyes (Fig. 51), and that will settle the

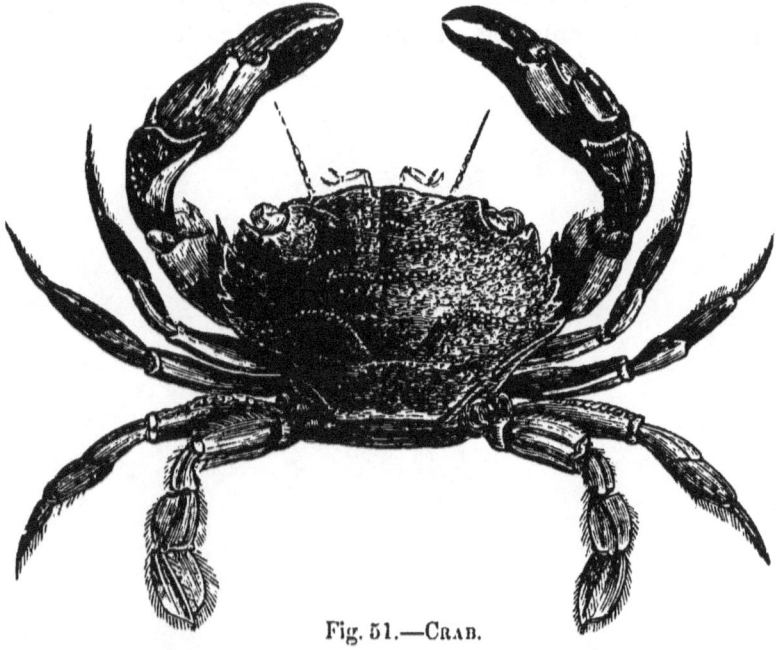

Fig. 51.—Crab.

question. Walking, as they do, forward, backward, and even sideways, with equal ease, it seems as if they too might be slightly puzzled about their formation, and so,

not stopping to decide which part is intended to go fore-
most, they dart off on a venture, and in the oddest manner
possible.

2. **Abundant on the Sea-shore.**—They are so abundant on
all our sea-shores that we rarely lift a bunch of sea-weed
or poke among the rubbish there without disturbing their
haunts, yet they scramble off and hide in the sand so quickly that we are not much wiser for their discovery.   Let us
pick up some cast-off shell and make a closer examination.

3. **Cephalo-thorax.**—The bodies of higher animals contain three principal cavities—the head, thorax, and abdomen.   In crabs, on the contrary, the head and thorax are
so closely united that we cannot distinguish them, and they
are covered by the same shell.   The proper name for a
head and thorax thus united is "ceph-a-lo-thorax."

4. **Two Principal Parts.**—A crab, consequently, has two
principal parts — the cephalo-thorax and the abdomen —
each containing a number of segments of its own.   To the
cephalo-thorax are attached five pairs of jointed legs.   The
front pair are much larger than the others, and form the
claws.   The abdomen consists of six segments ; but it is
small and inconspicuous, being folded under the cephalo-
thorax.

5. **Stalked Eyes.**—The compound eyes of crabs are on
long stalks, and they may be turned in different directions
or folded back into little grooves in the shell.

6. **The Gills.**—Crabs breathe by gills, which are leaf-
like plates so situated as to be readily bathed with water.
They contain a dense net-work of blood-vessels, through
whose thin walls the oxygen in the water finds its way
and mixes with the blood to purify it.   The crab's heart
consists of a single contractile sac.

7. How Crustaceans shed their Shells.—Crabs are often spoken of as crustaceans. The name will at once suggest animals having a hard crust. As this crust contains a number of pieces exactly fitted to one another, it has been compared to the armor worn by soldiers in olden times. The manner in which it is shed during the growth of the crab is curious and interesting. This hard shell never increases in size; therefore, as the crab grows its shell becomes too small, and is cast off. The discarded shell has the eye-stalks and legs attached, and looks like the perfect animal. When the proper time for this change arrives, the body shrinks away from the shell, separating from it at all points, and the animal works its way out of its old case. The exhausted creature now remains quietly in some secluded place, increasing rapidly in size, until the soft skin again hardens into a new shell.

8. This is a painful and perilous experience for the poor crabs. Occurring as it does several times in the summer, their weak and unprotected bodies fall an easy prey to their enemies, and they are often devoured even by other crabs which happen to be in better plight. Now it is that they are known as "soft-shelled" crabs.

9. Destroyed by other Animals.—Crustaceans, when fully coated with mail, are strong and destructive, fighting among themselves as well as with other animals. They eat any small creatures that come in their way, whether living or dead. On the other hand, they themselves are destroyed by larger animals, and crustaceans form a great part of the food of star-fishes, sea-urchins, mollusks, and many kinds of fishes and birds; consequently, great numbers of them are killed before reaching their full size. To protect the race from destruction by this loss of life, all crustaceans produce a large number of eggs.

10. **Young Crabs unlike the Old Ones.** — Young crabs (Fig. 52) are so unlike the full-grown ones that natural-ists formerly thought they belonged to a different class of animals. While yet very small they rise to the surface of the water, and swim about freely, until, after pass-ing through several changes, the body be-comes large and heavy towards the head, and

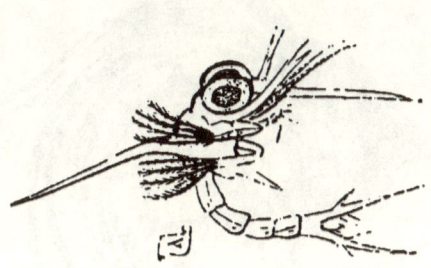

Fig. 52.—EARLY FORM OF THE CRAB.

the young crabs, losing the power of swimming, sink to the bottom, where they hide for a time. As they gain in size and strength, and are ready to begin their new manner of living, they creep towards the shore, and most of them pass the rest of their days in shallow water among the sea-weed.

11. **Where they Live.**—In the tropics some species live in the fresh water of brooks and rivers. Others live in the shades of damp forests; but, when breeding-time ar-rives, they visit the sea-shore to deposit their eggs. The land-crabs of Jamaica even live on the mountain-tops, yet every year they yield to a longing for their old home, and come down to the shores of the Caribbean Sea to lay their eggs. This duty performed, they return again to the mountains.

12. **The Hermit-crab.**—The hermit-crab (Fig. 53) is al-ways an object of interest. Unlike most other crustaceans, it has no shell to protect the soft body, and a tempting morsel is thus exposed. The hermit, conscious of its weak point, seeks shelter by taking possession of some spiral shell in which to place its soft abdomen. The hard claws

and the first two pairs of feet generally hang out over the edge of the shell, which henceforth moves about upon the

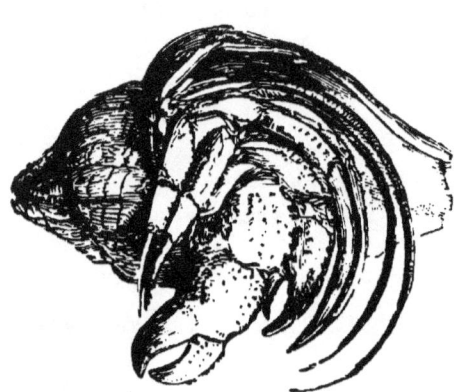

crab's back as if the two belonged together. The shorter hind-feet are roughened, thus enabling the crab to hold on to the inside of the shell, to which it clings so tightly that it will sometimes allow itself to be torn in pieces rather than quit its hold.

Fig. 53.—Hermit-crab.

13. As the hermit grows it needs to hunt up a larger home, and it may be seen creeping along the shore, examining and turning over shells to select one, often trying on several before it is suited—much as a boy might try on several pairs of boots before he is fitted exactly. Should a hermit fancy the shell of some living snail, it would not hesitate to kill and eat the owner, and then coolly take possession of the shell. Two hermits are sometimes found fighting for the same shell.

14. **Fiddler-crabs.**—Fiddler-crabs (Fig. 54) have one claw much larger than the other, and as they walk sideways they hold up the large claw in a threatening manner. They dig holes in the mud to live in, and they enter these homes with extreme caution. Running quickly to the entrance, they pause a while, turn their stalked eyes in every direction, and then slip suddenly in. They are not easily caught, for they dart into their holes quickly when alarmed.

15. **Effect of Use upon an Organ.**—The fiddler-crab is a good illustration of the effect of use upon any one organ.

The large claw so peculiar to this group belongs only to the males, who are great fighters. They use the large claw in their combats, which fact accounts for its increased size. The more peaceable females have no need of so powerful a weapon, consequently they do not possess this striking peculiarity of their mates, and, on the contrary, have small weak claws.

Fig. 54.—FIDDLER-CRAB.

16. **Pea-crabs.**—Many of you have seen the little round crabs that live in oyster-shells. These pea-crabs, or oyster-crabs, as they are called, are considered a delicacy, and are sometimes collected and sold by the dozen. Having no hard covering, they always take up their abode within the shell of the oyster or some other bivalve. They are not prisoners within the shell, as they venture out into the water sometimes, and return again when they wish.

17. They are said not to annoy the oyster in the least, or to deprive it of its food, since they eat certain small animals which float into the shell, but which the oyster never feeds

upon. Strange to say, it is only the female that shuts her-self up within the narrow limits of an oyster-shell. The male is much smaller, and frolics about on the surface of the sea.

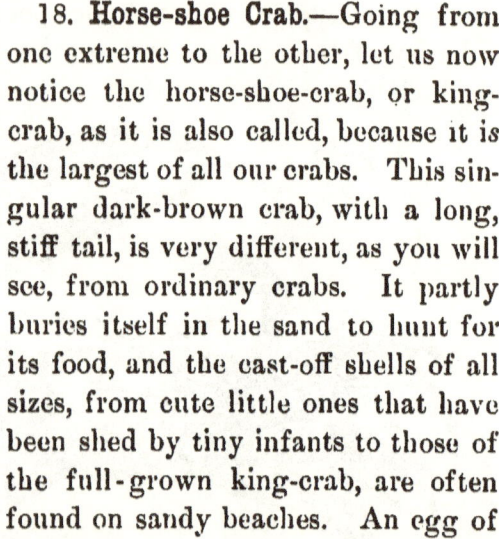

18. **Horse-shoe Crab.**—Going from one extreme to the other, let us now notice the horse-shoe-crab, or king-crab, as it is also called, because it is the largest of all our crabs. This singular dark-brown crab, with a long, stiff tail, is very different, as you will see, from ordinary crabs. It partly buries itself in the sand to hunt for its food, and the cast-off shells of all sizes, from cute little ones that have been shed by tiny infants to those of the full-grown king-crab, are often found on sandy beaches. An egg of

Fig. 55.—KING-CRAB

the king-crab, one-third larger than life, is shown in the illustration (Fig. 55). Several days before the egg hatches the young crab may be seen sporting about within the transparent shell of the egg.

XV.

LOBSTERS.

SUB-KINGDOM, ARTHROPODA : CLASS, CRUSTACEA.

1. Lobsters and Crabs Compared.—Lobsters, as well as crabs, have the head and thorax united, forming a cephalo-thorax. The compound eyes are on long, movable stalks. Behind these are two pairs of jointed antennæ or feelers. But near the mouth are five pairs of "jaw-feet," which we do not find in crabs; these, and some other additional organs, point to a difference in their manner of living which we shall now consider.

2. As lobsters live entirely under water, they breathe only by gills, which are richly supplied with blood. The gills are situated in a cavity under the body, and each plate of the gills is attached to one of the legs. A current of sea-water is kept passing over the gills, partly by the movement of the legs, and partly by a spoon-shaped appendage to the second pair of jaw-feet, which constantly bales out the water from the gill cavity.

3. The Claws.—A pair of true walking legs is attached to each of the last five segments of the cephalo-thorax. The front pair of legs forms the claws; these are very large, and are armed with strong pincers. One claw has sharp teeth for tearing food; the other has rounded teeth, and is used as an anchor to hold on to fixed objects. Lobsters are so quarrelsome that fishermen, before sending them to market, often fasten their claws open with plugs to prevent their fighting and injuring each other. Like some other animals we have studied, lobsters can throw off their own legs and claws when wounded or alarmed. New ones grow in their places, but we often see lobsters with limbs that do not match each other in size.

4. **The Swimmerets.**—Lobsters are great swimmers, and they are well suited to this favorite sport. The large abdomen consists of six segments, each bearing a pair of paddles called "swimmerets," while the body ends in a broad fin or "telson." Each appendage is attached to a segment of its own, and it is thought by some naturalists that the jaw-feet, antennæ, and eye-stalks are all modified limbs attached to their appropriate segments.

5. **How Animals are Fitted to the Life they Lead.**—It is interesting to notice how well every animal is fitted to the life it leads. The lobster by striking its powerful tail upon the water takes a long spring, and in this way the abdomen is a great assistance in swimming. The claws are also constantly used to catch the prey and to defend the animal, hence both the claws and the tail are large and muscular. These two organs we know contain the chief eatable part of the lobster. The crab, on the other hand, living upon the sandy bottom of the ocean, and among the rocks on the shore, is accustomed to walking or running, and as the abdomen is not needed for swimming purposes, it seems to have dwindled away to a mere apology for a tail, which is snugly tucked up beneath the cephalothorax. This same thing happens everywhere in nature. An organ that is no longer needed or used, shrivels in size and sometimes wholly disappears.

6. **Internal Organs.**—We must next know something of the internal organs of the lobster. A short œsophagus or gullet leads from the mouth to the stomach. The stomach

is very large, and contains curious pieces of cartilage to which strong grinding teeth are attached for crushing the food. These teeth are often called the "lady in the lobster." The large liver is of a dark-green color, and the heart consists of a single contractile sac, just back of the head.

Fig. 56.—LOBSTER (HOMARUS VULGARIS).

7. The Lobster outgrows its Shell.—The legs, antennæ, and even the eye-stalks of lobsters are incased in a hard shell, which, like the crab's shell, never increases in size; consequently, as the animal grows larger, its shell becomes too small. At such times the lobster loses its appetite, and hiding in some secluded corner, grows weak and thin. In this way the body shrinks from the shell. The shell then splits open, and after a good deal of screwing and twisting, the soft, tender body creeps from its outgrown case, drawing out with it legs, claws, eye-stalks, and all.

8. Forms a New Shell.—How weak and defenceless the

5

poor creature must feel with no coat of mail! It seems aware of the danger of being eaten by its hungry neighbors, so it remains out of sight. The lobster now fills its body with water and swells out as much as possible, and the soft skin, which is covered with a sort of glue, hardens to form a new shell, fitting tightly over every part just as the skin did.

9. **Color.**—The natural color of lobsters is a greenish-black, but it changes to a bright red in boiling. The females are sometimes found with large masses of eggs fastened to the swimmerets. The young ones are hatched with the same form as their parents, and they do not pass through the changes we noticed in young crabs.

10. **Manner of catching Lobsters.**—Lobsters choose the deep, clear water along rocky coasts for their dwelling-place, and on account of the delicacy of their flavor, they are caught in large numbers for food. A common way of catching them is by the use of "lobster pots." These are wooden cages with a funnel-shaped hole in the top, through which the lobsters can easily enter, but cannot so easily get out again. The pots are sunk in the water and marked by buoys. Great quantities of lobsters are caught on the coasts of the British Isles, and are often kept in perforated chests floating on the water until they are sent to market.

XVI.

BARNACLES.

SUB-KINGDOM, ARTHROPODA: CLASS, CRUSTACEA.

1. **Barnacles form a Coating on Rocks.**—Boys and girls who have been to rocky sea-coasts may have noticed a dull white coating upon the rocks after the tide has gone down. If they have given the subject much thought, they have probably discovered that on the cliffs this coating forms a strip reaching only to high-water mark.

2. At first we may think the rocks quite disfigured, but so great is the charm which living beings have for us that we shall become interested at once upon learning that this rusty covering consists of acorn-barnacles.

Fig. 57.—ACORN-BARNACLES.

3. **Shells closed at Low Tide.**—Any rocks that stand between high and low water mark may be chosen as the resting-place of these curious creatures. When the rock is left high and dry above the water there is nothing attractive about the barnacles. Their shells are tightly closed (as seen in Fig. 57), and they appear per-

fectly lifeless; but watch them when the tide comes in, and they will show signs of returning activity.

4. **Manner of Feeding.** — With the first welcome wave that reaches their resting - place you will see the valves within the acorn open, and a delicate feathery cluster of

arms will be thrown out of each barnacle, as in Fig. 58, and then suddenly disappear. This movement is repeated every few seconds with great regularity, and makes a current in the water, carrying towards the mouth small floating bodies on which the barnacle feeds.

5. The shell consists of two parts, one within another. The outer one is composed of several plates, open at the top; within it is a conical movable lid, the plates of which are opened and closed every time the arms are thrown out. In this way barnacles fish vigorously, as if they understood that two tides mean but two meals during the day, and consequently they must make the best use of them.

Fig. 58.—ACORN - BARNACLE, WITH ARMS EXTENDED.

6. This fishing is a graceful operation, and if you should find a large rock covered with barnacles, and bathed with clear sea-water, you will soon be fascinated with watching their motions. As the valves at the top of each cone open, twelve pairs of light, feathery arms are thrown out and drawn in again with unvarying precision.

7. **Young Barnacles more highly developed than Full-grown Ones.** — Young barnacles, when first hatched, are active, restless creatures, swimming about like young

crabs, but as they grow older they attach themselves to rocks, shells, drift-wood, sea-weed, sponges, turtles, or even to jelly-fishes. The head is firmly glued to these objects by a cement which the animal secretes. The rest of the body is free, and can be extended beyond the shell. Fig. 59 shows the body of a barnacle as it looks within the shell.

8. While young, and frolicking about in the water, barnacles have two well-developed eyes, but these dwindle away when the animal settles for life, and they finally disappear altogether. The shelly covering now grows, and henceforth barnacles are quiet, orderly individuals, never moving from the spot which they have chosen as a resting-place unless this happens to be upon a living animal or some

Fig. 59.— BODY OF GOOSE-BARNACLE.

floating object. So you see barnacles are really more highly developed in youth than they are later in life. Before growing into perfect barnacles they have parted with their sight, and with the power of moving or swimming from one place to another.

9. **Clinging to Vessels.**—Barnacles are found in all seas. They sometimes settle so thickly on the huge Greenland whale as to hide the color of its skin. They are also found clinging to the hulls of vessels in such large masses that the movement of the vessels through the water is retarded. These barnacles grow rapidly, and ships which start upon their voyages freshly painted have sometimes been obliged to put into port in order to have the barnacles scraped from the hull.

10. **Goose-barnacles.**—The goose-barnacle (Fig. 60) differs from the acorn-barnacle in hanging from a long mus-

cular stalk. The shell opens at the side, but the arrangement of the animal is the same as in the acorn-barnacle. It also has twelve pairs of jointed and ciliated limbs, which it throws out at regular intervals.

11. In former times these same goose-barnacles were thought to change into birds. There is a certain goose frequenting the western coasts of the British Isles, called the barnacle-goose, which was thought, even by learned men, to have sprung from the barnacle. The following quaint description of the transformation was written in the sixteenth century: " When the shell gapeth open " we see " the legs of the bird hanging out," then the bird, increasing in size, " hangeth only by the bill," and " in short space after it cometh to full maturity and falleth into the sea, where it gathereth feathers, and groweth to a fowl bigger than a mallard, and lesser than a goose." People believed that this change was actually going on before them, and there was some difficulty in proving it to be only a fable.

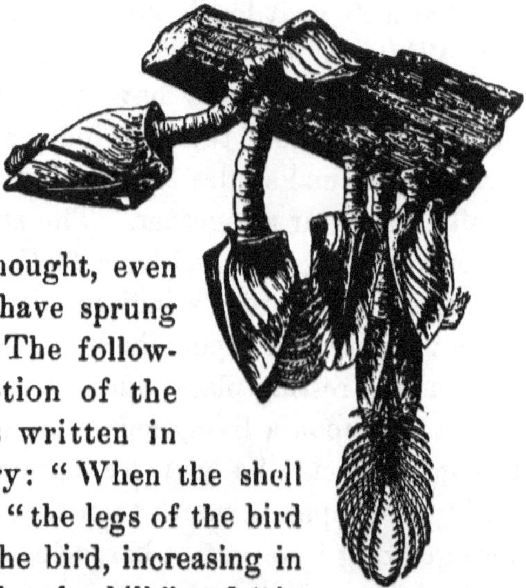

Fig. 60. — Goose-
BARNACLES.

XVII.
SPIDERS.

1. **Spiders.**—Although spiders are shunned and despised, they are mostly harmless creatures, quietly pursuing their work of destroying insects. They have a singular fancy for resting with their heads downward, and instead of living in pairs, they prefer to live alone. The females are usually larger than the males, and they show no good feeling towards their mates, eating them if they have an opportunity. In some cases, however, they live peaceably together for a time.

2. **Examination of a Spider.**—The two divisions of the spider's body are easily seen. The cephalo-thorax has a horny covering, but the abdomen is soft. It is entirely without limbs, and is united to the cephalo-thorax by a short stalk. Spiders have four pairs of legs, ending in hooks, which may be seen in Fig. 61. Near the mouth are hooked man-

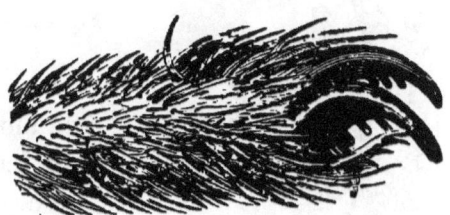

Fig. 61.—FOOT OF SPIDER, MAGNIFIED.

dibles, which contain a slit for throwing out a poisonous fluid to kill their prey. They have from six to eight eyes, which are grouped together on the top of the head.

Fig. 62.—Geometric Web of Garden-
spider.

The more highly developed spiders have a heart and blood-
vessels. They breathe by air-sacs and
tubes, which open on the under surface
of the abdomen. The nervous ganglia of
the head and thorax are united into a mass
slightly resembling a brain.

3. **Spinning Silk.**—Spiders are provided with

a curious set of machinery for spinning their webs. At the
end of the abdomen are three pairs of "spinnerets," the
last pair often extending behind the body like two prongs
(Fig. 63). On examining these spinnerets
we shall find them covered with tiny points;
from each of these flows a stream of gummy fluid, which hardens into silk when it
reaches the air. The movable spinnerets
are under the control of the spider, and
when they are held close together the fine
streams issuing from them unite into one
thread before hardening, but if the spinnerets are held apart the threads harden separately. By

Fig. 63. — Spin-
nerets of Spi-
der.

pressing the spinnerets against any object the fluid silk is
forced out of the tubes and adheres to its surface, consequently it is drawn out when the spinnerets are lifted.
The hind-legs are also used in helping to draw the delicate
stream of silk, and guiding it to form various designs.

4. In this way a spider's web which we can barely see
may be composed of more than a thousand threads. Like
a piece of ordinary rope, it is stronger for being made
up of small cords, but notwithstanding this the silk is too
delicate to be of service to man, and all attempts to weave
it into cloth have failed.

5. **Uses of the Web.**—Some spiders use their webs as traps
to catch their prey, and those that live in holes or underneath stones generally line their hiding-places with web.
Nearly all spiders enclose their eggs in a silken cocoon,
which, in some species, the mother carries on her back.
The young spiders remain in the web until they have grown
to a considerable size, when the mother sometimes tears
open the web, and the baby spiders may be seen swarming over her, as in Fig. 64. When the time arrives to

5*

wean them from her back the mother shakes or kicks them off with her feet, and they scamper away to begin life by themselves. Two thousand young spiders have been found in one cocoon.

Fig. 64.—FEMALE SPIDER WITH YOUNG ONES.

6. **Feeding the Little Ones.**—When feeding her babies, the mother holds a nice plump fly, or some such dainty morsel, while the little ones gather round and suck its juices. When their hunger is satisfied they run off, and a new set comes to the feast. The mother often kills some of her own little ones to feed the remaining spiders of her numerous brood.

7. **Cobwebs.**—What could be more charming than the filmy cobwebs that ornament the country road-sides, the fences, and the bushes in the early mornings of summer, every thread bearing a precious load of dew-drops? Although the webs remain through the day, they please us most when sparkling with dew. Those flat webs that are so familiar to us all slope down into a cunning little tube which leads off among the grass. If you look closely you will find the spider hiding just inside this tube, and watch-

ing intently for some insect to alight on its snare. When this happy event occurs, the spider runs out, and seizing its prey, carries it into the tube, where it sucks the juices of its victim and casts away the dead body.

8. **Garden-spider.**—Our common black and yellow garden-spider weaves a wheel-shaped web like that which is shown in the picture on page 104. This web is really a work of art. First, the framework for the wheel is made by a number of threads crossing each other at one point, and firmly attached at both ends to surrounding objects. These threads are like the spokes of a wheel, and upon them the spider fastens a spiral thread, making circle after circle.

9. The spider then stations itself, head downward, at the centre of the web, from which point it can feel the slightest motion made by an insect alighting upon it, and can quickly reach the spot, to secure its victim by additional threads. This garden-spider places her eggs in a pear-shaped cocoon, which you will see represented in the picture.

10. **Gossamer-spiders.**—Some small spiders have a fantastic habit of weaving balloons for themselves and sailing through the air. They pass by the general name of "gossamer-spiders." Placing themselves in some high position, such as the tops of fences, with their heads towards the wind, and their spinnerets open, they allow a stream of fluid silk to be blown out by the current. The spider then makes a spring, and, grasping the thread with its feet, is carried by the wind for long distances, completely surrounded by a mass of its own web. These little fairy balloons may be seen floating through the air almost any fine day in the autumn.

11. **Water-spider.**—Besides those that mount into the air, there are some spiders that live in the water. The

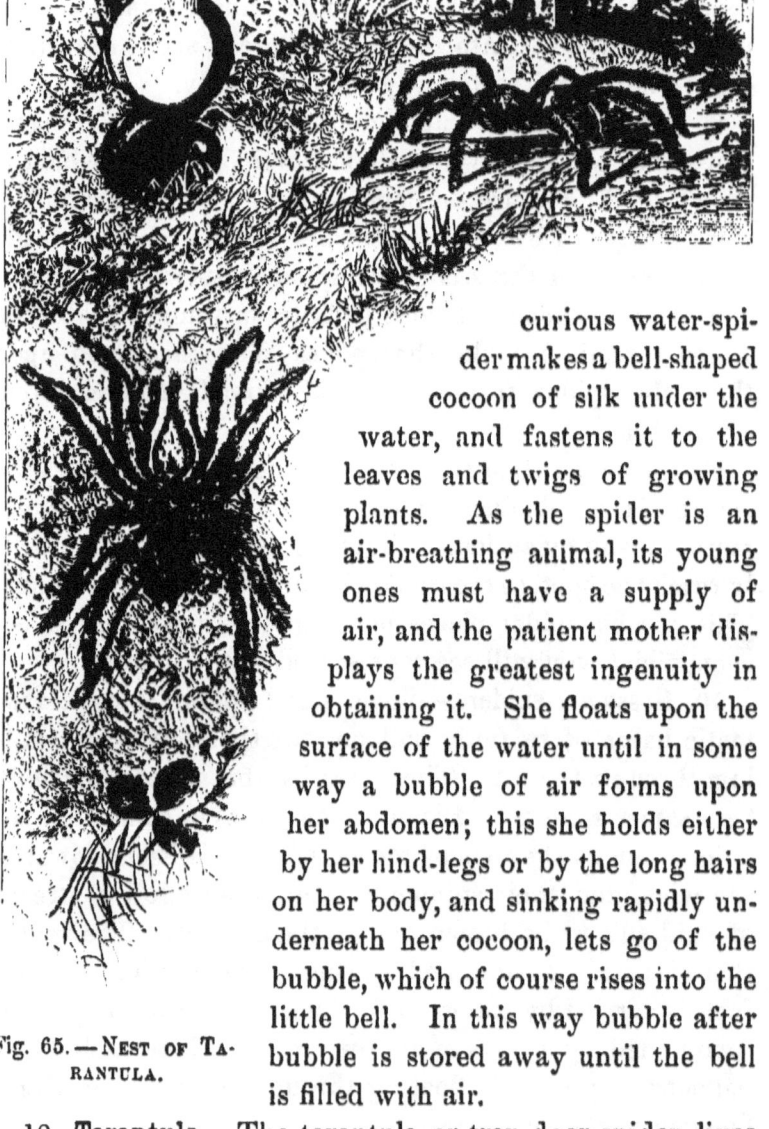

curious water-spider makes a bell-shaped cocoon of silk under the water, and fastens it to the leaves and twigs of growing plants. As the spider is an air-breathing animal, its young ones must have a supply of air, and the patient mother displays the greatest ingenuity in obtaining it. She floats upon the surface of the water until in some way a bubble of air forms upon her abdomen; this she holds either by her hind-legs or by the long hairs on her body, and sinking rapidly underneath her cocoon, lets go of the bubble, which of course rises into the little bell. In this way bubble after bubble is stored away until the bell is filled with air.

Fig. 65.—NEST OF TA-RANTULA.

12. **Tarantula.**—The tarantula, or trap-door spider, lives in warm countries, and digs for its nest a hole in the ground two inches or more in depth. The hole is neatly

lined with real raw silk, and tightly covered with a most ingenious lid. How do you suppose the spider manages to make this circular lid of the exact size, and then fasten it on with a silken hinge? The top of the nest is first covered with a web of the proper shape, on which is placed a small quantity of earth; over this is spread another web, then more clay, so that the lid is composed of layer after layer of web and fine clay, which harden into a thin, stiff mass. The webs on one side are attached to the edge of the nest to form the hinge.

13. If the lid is closed it looks so exactly like the surrounding soil that these nests are not easily found. The concealment is completed by a cunning habit of covering the door with moss, or some substance similar to that which grows around it. When in its nest the spider holds on to the door so tightly by its mandibles and fore-feet that the lid cannot be raised from the outside.

XVIII.

INSECTS.

SUB-KINGDOM, ARTHROPODA : CLASS, INSECTA.

1. Largest Class in the Animal Kingdom.—Insects themselves are mostly small, but the class to which they belong is the largest class in the animal kingdom, and contains more than two hundred thousand species.

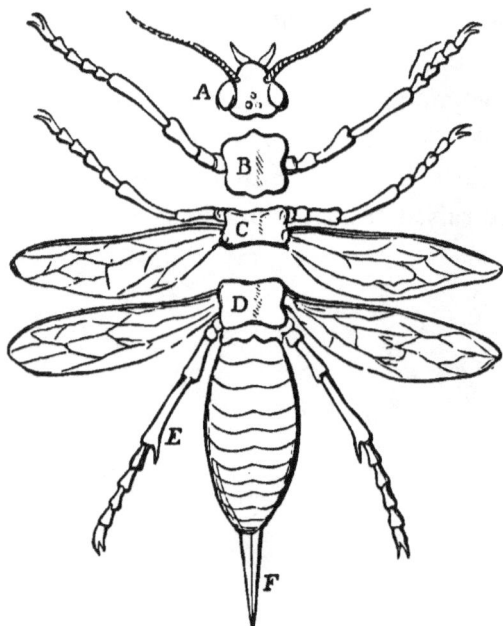

Fig. 66.—Diagram of an Insect.

A, the head; B, C, D, segments of the thorax; E, abdomen; F, ovipositor.

2. Found in every Part of the World.—These little creatures are found in all the countries and oceanic islands that man has reached; they inhabit hot springs as well as the coldest streams.   Humboldt found them on the Andes far above the line of perpetual snow, and Darwin, on the early voyage of the *Beagle,* found a dragon-fly two hundred and fifty miles from land.

3. **Body made up of Segments with no Skeleton.**—Insects have no internal skeleton, but they are covered with a horny skin. The head, thorax, and abdomen are entirely distinct, and each part is mostly divided into segments such as are represented in the diagram (Fig. 66).

4. **Compound Eyes.**—Conspicuous upon the sides of the head are the large round eyes, which, examined through a microscope, will be found covered with numerous flat surfaces or lenses (Fig. 67). These are called compound eyes, for they consist of a great number of eyes crowded into one mass; and they have the power of looking in many directions at the same time. In addition to their compound eyes, most insects have three simple eyes placed between them. The antennæ, or feelers, are also interesting, and you will find great variety in their shapes.

Fig. 67. — HEAD OF A BEE, SHOWING COMPOUND EYES, SIMPLE EYES, AND ANTENNÆ.

5. **Limbs of Insects.**—To the thorax are attached three pairs of legs, and mostly two pairs of wings. These wings are thin and delicate, and are very large in proportion to the body. They are supported by a net-work of hollow tubes which enclose air-pipes and blood-vessels side by side.

6. The abdomen has no limbs, and it often ends in a piercer or sting, which is called the "ovipositor." You may have noticed in larger insects a curious sliding in and out of the segments of the abdomen. This bellows-like action helps to change the air in the air-tubes.

7. **Organs of Digestion and Circulation.**—The œsophagus leads into a crop from which the food enters the gizzard, where it is crushed and passed on to the true stomach

Fig. 68.—ALIMENTARY CANAL OF A BEETLE.
b, œsophagus; c, crop; d, gizzard; e, stomach; g, intestine.

(Fig. 68). Insects have no distinct heart, and the blood is propelled by the contraction of eight sacs, which allow it to flow only towards the head. The blood is colorless, and fills the irregular spaces left between the organs.

8. **Breathing Apparatus.** — Insects breathe by tracheæ, which are air-tubes passing through every part of the body. Being filled with air, the tracheæ supply the blood abundantly with oxygen, and at the same time diminish the weight of the body. These tubes are composed of elastic threads wound in a close spiral (Fig. 69), giving them great strength and lightness, and preventing the possibility of their being pressed together and closed. The tracheæ open on the surface of the body in small holes, called "stigmata," which are arranged on the sides of the thorax and abdomen, and are so contrived as to admit air freely, while they exclude water or dust. A drop of oil on the abdomen of an insect will kill it by closing the stigmata and causing suffocation.

9. **Insects have no Voices.**—No insect is known to have a voice. The various noises of insects, so commonly heard, are caused by the rapid vibration of their wings, or by rubbing together some hard parts of their bodies.

10. **Metamorphosis.**—Most young insects

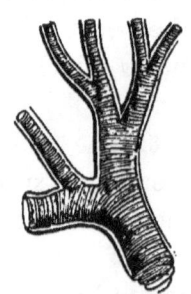

Fig. 69.—TRA-CHEÆ OF AN IN-SECT, SHOWING ELASTIC SPIRAL THREAD.

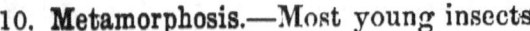

are very different from their parents, and before reaching their perfect state they pass through a succession of changes called "metamorphosis." As butterflies are familiar insects, let us take them for an example, and study the changes through which they pass.

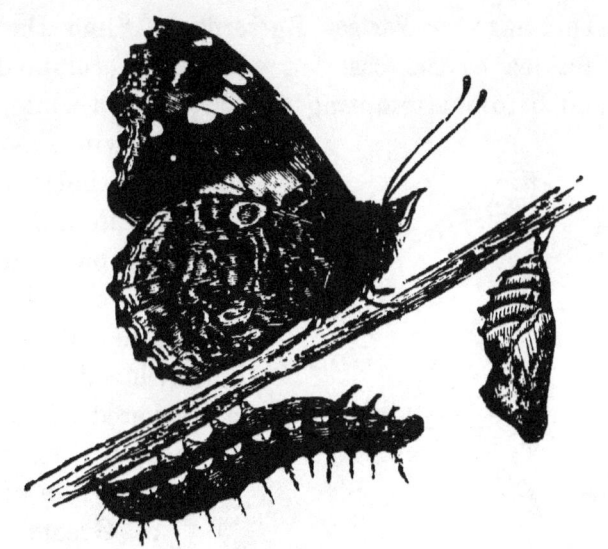

Fig. 70.—BUTTERFLY IN THE LARVA, PUPA, AND IMAGO STATE.

11. **The Larva of a Butterfly.**—From the eggs of butterflies are hatched young caterpillars. The caterpillar crawls over the plant upon which it was born, eagerly devouring the green leaves, as its mouth is fitted for chewing. It grows rapidly, and sheds its coat several times. During this period of its existence it is called a "larva."

12. **The Pupa or Chrysalis State.**—At length the larva leaves off eating, and enters the "pupa" or "chrysalis" state. Wrapped in a dry skin, and hanging head downward suspended by a silken thread, it remains for a time apparently dead. Shut up, however, in the silence of this

temporary prison, a marvellous change is going on, and when the skin bursts a full-grown butterfly appears, furnished with wings and arrayed in bright colors. These three stages are represented in Fig. 70. The attractive insect now in no way reminds us of the caterpillar from which it sprang.

13. **The Imago or Perfect Butterfly.** — When the butterfly first leaves the case its wings are crumpled and moist, and before attempting to fly it rests a while, until

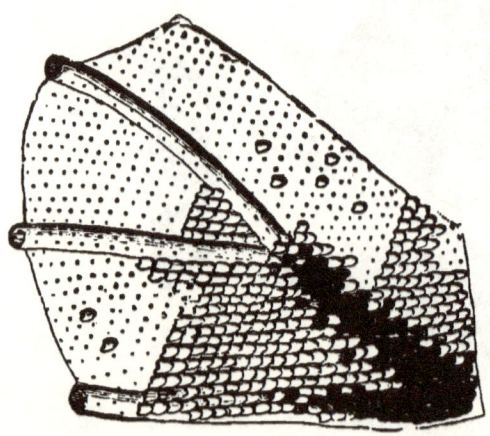

the wings stretch out to their full size. The delicate hues of the butterflies are due to the small feather-like scales with which they are covered. The scales overlap each other, as shown in Fig. 71.

14. Great changes have also taken place in the mouth, and henceforth a butter-

Fig. 71.—Scales on the Wing of a Moth.

fly sucks the sweet juices of flowers through a slender tube, which, when not in use, may be rolled up spirally under the head. Our beautiful insect has now reached the "imago" or perfect state, and the great aim of this part of its existence is to choose a mate. In this it makes no mistakes. The image of its own kind seems to be impressed upon its fancy, so that it never mates with any but its own species. Insects know each other when they meet, just as they know the right flowers to feed upon, and in the same way the female butterfly selects the proper spot for her eggs, generally placing them on some plant whose leaves are suitable food for her caterpillar children.

Fig. 72.—BUTTERFLIES.

15. Nearly all insects pass through these three conditions, the larva, the pupa, and the imago, as we have before stated. Their larvæ are known by the various names of caterpillars, grubs, and maggots. By keeping

a few caterpillars you may watch for yourselves all these interesting changes.

16. What could possibly seem more aimless than the joyous, careless flitting of a butterfly! Floating hither and thither through the bright sunshine, and folding together its elegant wings above the choicest flowers, its life appears a most luxurious one; still it has its own part to play.

17. **Flowers fertilized by Insects.**— It is a well - known fact that most bright-colored flowers are dependent upon the visits of insects (especially of bees) to perfect their seeds, and thus to keep up a succession of new plants from one year to another. The insects are attracted by the showy petals, and they enter the flowers to obtain the honey which is stored up in the bottom of the tube. In so doing, grains of pollen adhere to their heads and wings, and are carried from one flower to another. These pollen grains lodge upon the moist surface of the pistils as the insects brush past them, and in due time seeds are produced.

18. **Bright-colored Insects attracted by bright Flowers.**— Butterflies are great rovers. Having no homes of their own, they flit gayly about and visit the most brilliant blossoms. Throughout nature we find highly colored birds and insects have the same preference for bright flowers and fruits as is shown by the butterflies.

19. **Butterflies and Moths contrasted.**—Many of our moths resemble butterflies; and as both of these insects change from caterpillars, it will be well to notice some of the differences between them. In the first place, true butterflies fly only in the daytime. Their antennæ are long and thread - like, with knobs at the end. When at rest, the wings are generally folded together and held erect above

the body, thus concealing the more brightly colored upper surfaces, and affording the insect protection against its enemies. The under side of the wings often resembles in color the flower upon which the butterfly feeds.

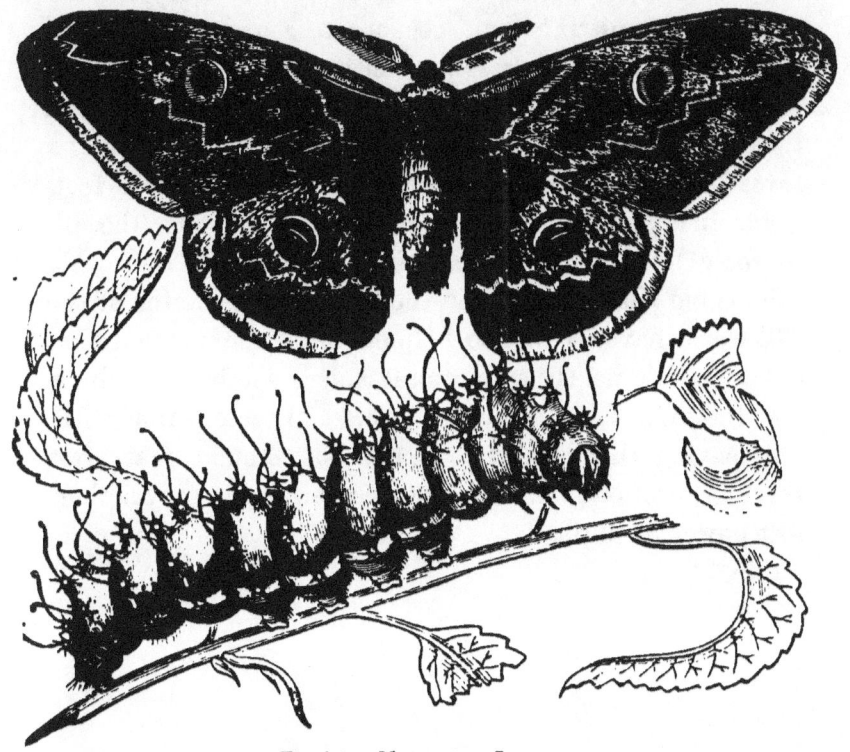

Fig. 73.—MOTH AND LARVÆ.

20. Moths fly only at night or during twilight. The body is generally stouter and more robust than that of the butterfly. Their antennæ are tapering, and sometimes beautifully feathered (Fig. 73). They do not fold their wings in repose, and their larvæ enclose themselves in silken cocoons.

21. **Silk-worms.**—Silk-worms, the most useful of these

insects, are extensively cultivated for the silk of their co-coons. When the pupæ are ready to leave the cocoon they make a hole through it for their escape, which breaks the thread of silk. To prevent this it is custom-ary, when silk-worms are raised for profit, to kill the pupæ by submitting their cocoons to a great heat. The cocoons are afterwards soaked in warm water to soften a gummy substance which they contain, and the silk can then be wound off in an unbroken thread. The length of a thread of silk has been estimated to be nine hundred feet.

22. In commencing its cocoon the larva attaches the silk to some fixed object, then winds itself in its own web, thickening the cocoon upon the inside. The moths of the silk-worm have grown so helpless from confinement that the female is nearly as motionless as if she had no wings, and the male merely flutters around his companion with-out leaving the ground. It has been found that after three generations raised in the open air they recover their lost power of flight.

XIX.

BEES.

1. Transparent Wings Hooked.—You may have noticed how thin and transparent the wings of bees are, and that they are supported by delicate veins. Look at them now with your microscopes, and you will see small hooks on the edge (Fig. 74), which fasten together the front and back wing during flight, in order that they may move as one wing.

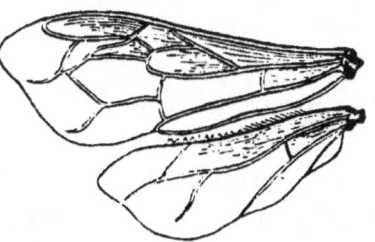

2. The Sting.—The sting of the female (Fig. 75) is a remarkable instrument at the end of the abdomen.

Fig. 74.—WINGS OF A BEE, SHOWING THE HOOKS.

It consists of two darts, *a*, and a sheath, *b*, connected with a poison - gland, *c*. The wound is first made with the sheath, after which the darts are thrust out to deepen it. These darts have a number of pointed barbs at the end, *d*, and it is difficult to remove them from the wound, so they sometimes break off. This loss of the sting causes the bee to die, though not always immediately. The sting, or ovipositor, varies in form with different kinds of bees, and it is sometimes used for cutting, boring, and sawing holes in which to deposit the eggs. Male bees have no sting, and are therefore harmless.

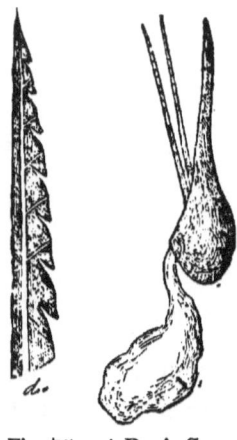

Fig. 75.—A BEE'S STING.
d, dart magnified.

3. **Social Bees and Solitary Bees.**—Humble-bees and hive-bees live in large families, and are called social bees. There are also solitary bees which live entirely alone.

4. **The Carpenter-bee.**—The carpenter-bee is an interesting example of a solitary bee. She bores her nest in old wood, mostly selecting the dead limb of a tree, an old post, or wooden railing. One of these nests is shown in Fig. 76. The bee bores a tube which soon makes a sudden turn, and is continued several inches down the trunk, parallel to the grain of the wood. This tunnel is afterwards divided into cells, in each of which is placed an egg with a supply of food for the young larva. The partitions between the cells are made of the sawdust which has collected from her boring, moistened with a gummy fluid which the bee secretes. She

Fig. 76.—NEST OF CARPENTER-BEE.

seems to know that the egg first deposited at the bottom of the tube will hatch first, so she bores a second opening at that part of the tunnel, through which the young bees come forth in succession at the proper time.

5. **Humble-bees.**—Humble-bees, as we have said, are among the social bees. They make their nests in holes in the ground (Fig. 77), often taking possession of a deserted mouse nest. All the colony, except the females, die when winter comes. These females remain in a torpid state, concealed among moss or rotten wood, to start new colonies the following spring.

6. **Hive-bees.**—The habits of hive-bees are exceedingly curious, and deserve our especial study. Every hive contains a queen-bee, workers, and drones (Fig. 78).

7. **The Workers Build the Nests.**—The whole

Fig. 77.—NEST OF HUMBLE-BEE.

labor of building the nest and providing for the large family falls upon the workers. They have a softer material to work in than the carpenter-bee, since their nest is built of wax, which is a secretion of their bodies, and which

6

forms in scales between the segments of the abdomen. With their feet the bees remove the wax, and work it with

Fig. 78.—HIVE-BEES.

a, queen; *b*, worker; *c*, drone.

their mouths and mandibles, mixing it with saliva until it becomes soft and white.

8. It is then placed upon the ceiling of the hive, and the cells are carefully shaped and fitted to each other, forming the honey-comb which is our wonder and admiration. The manner in which the six-sided cells fit together gives the greatest possible amount of space, while it requires the least material for building.

9. **Gathering Honey.**—In collecting honey for the hive a bee goes steadily from one blossom to another, visiting flowers of only one kind on each excursion; thus it does not mix the honey from different flowers. The long tongue, or proboscis, enters the tube of the flower and laps up the honey. The tube of some flowers is too long and narrow for the bee to enter, so the honey is sucked

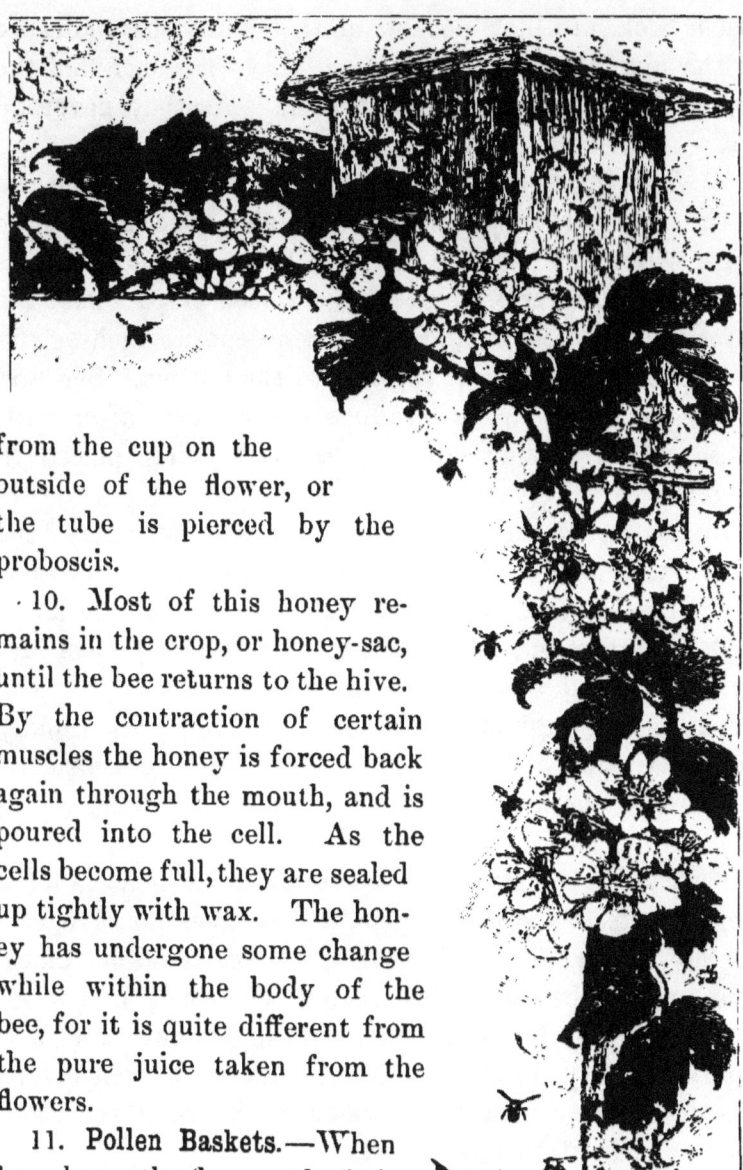

from the cup on the outside of the flower, or the tube is pierced by the proboscis.

· 10. Most of this honey remains in the crop, or honey-sac, until the bee returns to the hive. By the contraction of certain muscles the honey is forced back again through the mouth, and is poured into the cell. As the cells become full, they are sealed up tightly with wax. The honey has undergone some change while within the body of the bee, for it is quite different from the pure juice taken from the flowers.

11. **Pollen Baskets.**—When bees leave the flowers the hair on their bodies and legs is covered with pollen, which they

Fig. 79.—LITTLE PLUNDERERS.

brush back into little pockets on their hind-legs and carry to the hive. It is a singular fact that the queen and the drones have no such pollen baskets. As they never go out to gather honey, they need none.

12. **The Queen-bee.**—Each hive has one queen, and she is the only perfectly developed female. She lays all the eggs, which sometimes amount to two thousand in a single day. Different sized cells have been prepared for the three classes of bees, and the queen deposits each egg in its proper cell, gluing it slightly to the bottom. She first lays eggs which are to produce the workers, afterwards those which produce drones, the last being placed in larger cells.

13. **Duties of the Nurses.**—In three or four days the eggs hatch into little white grubs, and then the duties of the nurses, or workers, begin. The nurses feed the larvæ with a mixture of pollen and honey, which they have first swallowed, and which is already partly digested. The larvæ require a great quantity of food, and grow rapidly until they almost fill the cell. When they refuse to eat any longer, the nurses seal over the cells until the young bees are perfectly developed.

14. **The Perfect Bee.**—Fastened within its cell, the larva spins for itself a silken cocoon, and remains inactive, eating no food while the wonderful change is taking place. The care of the nurses has ceased, and when the perfect bee is ready to leave the cell it struggles out alone, and enters the busy throng outside with no one to welcome it. The workers soon take possession of the empty cell, and prepare it for future occupants.

15. **Treatment of a Young Queen.**—On the other hand, the young queen in her cell is treated with the greatest distinction. The larva is given richer food and in larger

quantities than the workers or drones receive. When she is ready to leave the cell, the workers gather around and gnaw at the top of the cell until it is so thin that the movements of the young queen within may be watched. A hole is made in this cover large enough for her to extend her proboscis, and she is fed in this position for several days, uttering the while a peculiar cry called piping.

16. The queen seems to have a hatred for those of her own sex, and she will destroy the young queens that come within her reach. Consequently, if the bees have not yet swarmed, the workers do not allow a young queen to stir from her cell. After the old queen has left the hive with her swarm, the young queens are liberated at intervals of a few days, and they lose no opportunity to kill each other.

17. **The Larva of a Worker may develop into a Queen.**— If by any accident the hive is left without a queen, the bees are thrown into great excitement, but they soon waken up to the necessity for action, and they begin, as it were, to cultivate a queen. They select three adjoining worker cells which contain larvæ, and cutting away the partition walls, convert them into one large cell. Two of the larvæ are destroyed, and the remaining one, by being fed on royal food, and having plenty of room and other favorable conditions, grows into a queen instead of a worker. This slight change of treatment not only gives her a different form and color, but it alters her whole nature, and gives her different instincts.

18. So you will see that queen-bees and workers come from the same kind of larvæ, and that these larvæ develop, according to the circumstances under which they are placed, either into queens or into workers.

19. **Drones killed by the Workers.**—The drones are

males, and they take no part in the work of the hive. In the latter part of summer the workers kill them without mercy, as if they were determined to support them no longer. They attack the drones, and sting them between the rings of the abdomen, afterwards throwing them out of the hive.

20. **Swarming.**—Bees usually swarm, or fly off in search of a new home, in the spring, never leaving the hive, however, until it is well stocked with eggs and the weather is warm. When about to swarm, the queen and the workers become very much agitated, hurrying to and fro for several days before they start. As the time for departure arrives, several bees fly in circles around the hive; suddenly the noise and bustle are hushed, and they all enter within. At a given signal, those which are to compose the swarm fly off rapidly, and select some tree or bush on which to alight. If their queen is not with them, they soon discover the mistake and return to the hive, where they wait for several days before a second attempt is made.

21. When the bees have entered their new home, they arrange themselves in a loop, or festoon, by hooking their claws together, and in this manner they hang from the roof of the hive. Thus they continue motionless for some time, while a store of wax is forming with which to build their new comb.

22. The bees which remain in the old hive after the swarm has left quietly pursue their labors, and a new brood soon fills the vacancies. The young queens, in their turn, lead off new swarms, and thus proceeds the busy life in a beehive. There are sometimes as many as fifty thousand bees in one hive, yet the work goes on without the slightest disorder or confusion.

23. Ventilating the Hive.—The workers keep the hive perfectly clean, and allow no dead bees or other impurities to remain within it. They are also careful that it shall be well ventilated. To accomplish this a certain number of bees continually fan their wings as if flying, although their feet are fastened to the floor. Some bees are occasionally stationed outside the hive to perform the same movements, but the greater number are within, one set relieving another after a certain time. The rapid motion of their wings causes a current of fresh air to pass through the hive; it also produces the humming sound which is constantly heard from a hive of bees.

XX.

WASPS AND MOSQUITOES.

SUB-KINGDOM, ARTHROPODA: CLASS, INSECTA.

1. **Wasps.**—Wasps have a general resemblance to bees, although they may be distinguished by their wings, which, when at rest, are laid over the body; also by the deep stalk-like division between the thorax and abdomen.

Wasps differ greatly in their habits. Like the bees, some live alone, others live in colonies.

2. **The Mud-wasp.** —Our common mud-wasp is among the solitary ones. This wasp makes its nest of mud, fastened to the side of a wall or under a ceiling. The nest consists of long cells arranged horizontally. In each

Fig. 80.—Digger-wasp—Cocoon and Larva.

cell is deposited one egg and a supply of little spiders for the young larva to feed upon after it is hatched. The spiders are not killed, but only stunned, and imprisoned alive when the end of the cell is fastened up.

3. In Fig. 81 you see a cell which has not yet been closed. The remaining cells were full of little green spiders, still active, when this nest was found.

4. **Social Wasps.** — Social wasps live in large families, which contain females, workers, and males. When winter approaches, all the wasps die except the females; these creep into some safe place, and sleep through the cold weather with their wings and legs tightly folded. In

Fig. 81.—NEST OF MUD-WASP.

the spring they revive, and each female starts a new nest for herself.

5. **Nests built of Paper.**—The nests of social wasps are always built of paper. Indeed, wasps were the first paper-makers. Long before man had learned the various processes required for manufacturing it, wasps had mastered the secret. Their paper is beautifully variegated, and being made of the fibres of wood, it is so durable as to bear exposure to rains and storms. Gnawing these fibres from some old fence or tree-trunk, the wasps moisten them with saliva until by the action of their jaws a paste is formed ready to spread out in a thin sheet. In looking at a piece of this paper, the wavy stripes will show just how far each bundle of fibre went towards forming the nest.

6. **Starting the Colony.**—As we have stated, there is but one wasp to do all the work in starting the home, so the building goes on slowly at first. By the time three or

6*

Fig. 82.—NESTS OF
SOCIAL WASPS.

four cells are fin-
ished the young
workers which
occupied them
are ready to help
the mother, who has been
busy building the nest, depositing eggs, and feeding the hungry
larvæ. Other cells are made and more
eggs deposited, the work going on
rapidly. The first wasps hatched are
the workers; the perfect males and
females do not appear until nearly the end of
the season.

7. Some kinds of wasps make their nests in holes in the ground, others fasten them to walls or to the branches of trees. The flat nests in Fig. 82 are built without any covering to the cells.

8. **The Hornet's Nest.**—A much more elaborate nest is made by the hornet. The one represented in Fig. 83 is

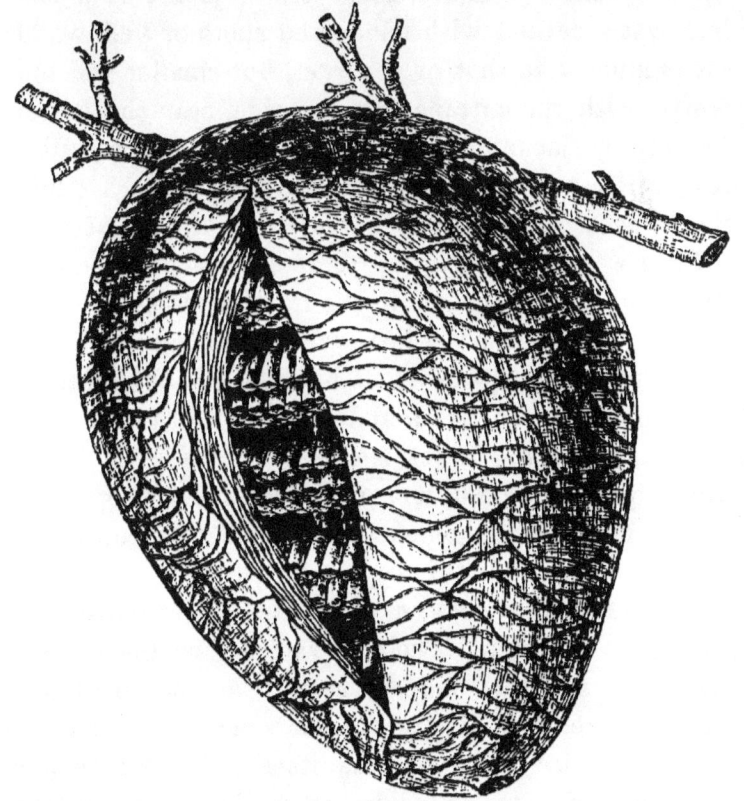

Fig. 83.—HORNETS' NEST.

cut open at one side to show the interior. It is formed of tiers of cells, one above another, with their mouths open-ing downward; the tiers are attached to little stalks which hang from the top of the nest. The whole is covered with several envelopes of paper, and the entrance is through a circular opening in the bottom. When it becomes necessary to enlarge the nest, new envelopes are added on the outside, and the inner covers are removed to make room for more cells. These nests are found in the woods, attached to the branches of trees.

9. **The Yellow-jacket.** — The yellow-jacket is a small black wasp marked with bands and spots of yellow. Its nest is much like that of a hornet, but smaller and more pointed, with the entrance on one side, near the bottom. The yellow-jacket sometimes attacks persons without provocation, and its sting is very severe. As a general thing, wasps do not sting unless they are irritated, but they are zealous in guarding their nests, and become agitated upon any approach to it ; if it is molested in any way, they rush upon the intruder without mercy.

10. **The Mosquito.**—Now let us glance at the mosquito. Its sting is on the head, and consists of several sharp lances and sucking tubes enclosed in a flexible sheath. After lancing the flesh, mosquitoes let fall a drop of poison, which makes the blood so thin that they can readily suck it through their tubes.

11. **Young Mosquitoes in the Water.** — Mosquitoes lay their eggs in water. Their larvæ pass by the name of "wigglers," and they may be seen in any stagnant pool. Here they remain during winter, when the ponds are covered with ice, and the mosquitoes of last season have been killed off with the cold. So, while we are enjoying a rest from the attentions of these little pests, another generation is coming on for next season.

12. The larvæ move through the water by sudden jerks. Their breathing organs are towards the tail, so they swim with the head down, as may be seen at E in Fig. 84, but

after they throw off the first skin and enter the pupa state, they breathe through the thorax, and keep the head at the surface of the water. Once more the skin splits, and they fly away full-grown mosquitoes. The dry case of the pupa forms a sort of boat, upon which the insect may rest and spread its wings before taking flight.

13. You may see this interesting metamorphosis going on in any pond in summer-time. A bright sunny morning brings thousands of these little boats to the surface, and you may be so fortunate as to see the occupant burst its shell and fly off into the sunlight.

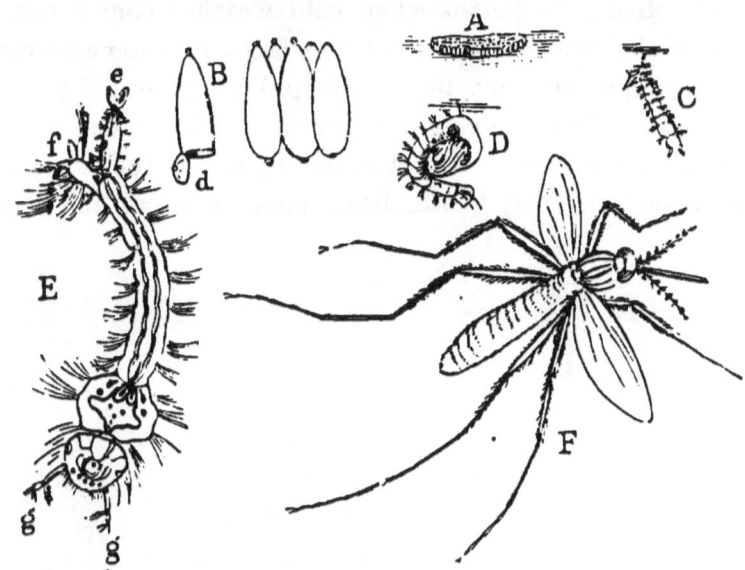

Fig. 84.—DIFFERENT STAGES IN THE GROWTH OF A MOSQUITO.

A. boat of eggs; B, eggs highly magnified; d, with lid open for the escape of the larva; C, D, pupæ; E, larva magnified, showing respiratory tubes (e), anal fins (f), antennæ (g); F, imago.

14. **The Eggs.**—The eggs of mosquitoes are cemented together side by side, and protected by some water-proof covering which enables them to float securely upon the

water, like miniature life-boats, which they really are. Each egg, moreover, contains a tiny air-bubble, and if the little life-boat should happen to be plunged beneath the surface, it rises quickly, and always with the right side up. These rafts of eggs are shown nicely at *A*, in Fig. 84. At *B* you will see the eggs magnified, with a curious lid at *d*, for the escape of the larva.

15. **House-flies.**—Our common house-flies live with us on intimate terms, and take great liberties in our homes; still, the early part of their lives is concealed from us, and we scarcely think about how they come or where they go.

16. Most flies perish when cold weather comes, but a few of the strong, healthy females creep into some crevice or corner, where they lie in a torpid state until the next summer. Here the eggs are deposited from which a new generation springs. In hot climates, and in rooms which are kept constantly warm, flies remain active all the year.

XXI.

ANTS.

SUB-KINGDOM, ARTHROPODA: CLASS, INSECTA.

1. **Remarkable Instinct of Ants.**—Ants are considered the most highly developed of all insects. Indeed, none of the lower animals possess such remarkable instincts as the ants. They show great wisdom and ingenuity in building their nests and in reaching any desired point. They make roads for themselves by carefully removing any obstacle in their way. They also dig tunnels of considerable length, sometimes resorting to this method for crossing broad rivers. They protect their nests, fight battles, gather food, tend their young, take care of domestic animals, and possess slaves. Their industry is not excelled by the bees and wasps. They work all day, and, when there is necessity, even at night.

2. **Their one Flight into the Air.**—Ants live in families, consisting of males, females, and workers. At first the young males and females are furnished with wings, and they fly from the nest to select their mates. Immediately after this first and only flight the males die, and the females strip off their wings, and do not leave the nest again.

3. **Labor of the Workers.**—The workers are much more numerous than the other classes; some of them serve as soldiers, others, which are generally smaller, serve as

nurses. All the labor of the colony falls upon the workers, and they attend to their various duties in the most
orderly manner.

4. **Ants' Nests.**—Ants do not all build their nests in
the same way. Some species heap up a mass of small
sticks and pine leaves; some bore into the trunks of old

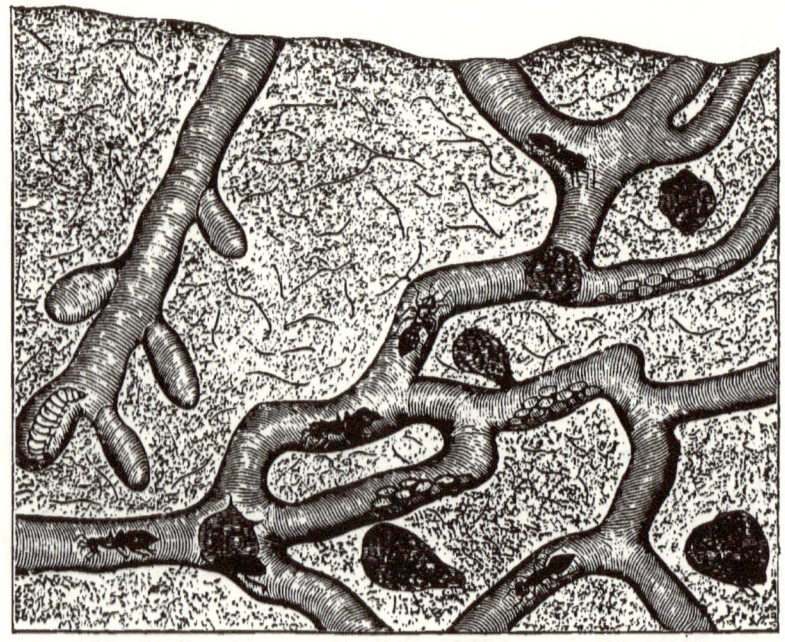

Fig. 85.—ANT NEST, WITH UNDERGROUND PASSAGES.

trees; but most ants make holes in the ground, with a
little mound of earth around the entrance, which we call
an ant-hill. These nests are carefully contrived, with
passages and avenues leading to many chambers, as you
will see in Fig. 85. The entrances are closed every night,
and opened in the morning. If it rains during the day
they remain closed, and the ants are confined within the
nest.

5. **Nurses' Care of the Eggs and Grubs.**—The eggs, which are scarcely large enough to be visible, are not deposited in any especial place by the females, but are immediately taken possession of by the nurses, who carry them to some favorable place, and who are henceforth devoted in their attentions to them, constantly licking and cleaning them, and frequently changing their position.

6. From the eggs are hatched little white grubs, which are entirely dependent upon their nurses for food. Every morning they are carried into the sunshine, or at least to the upper chambers that have been warmed by the sun, and towards evening they are again taken back to the bottom of the nest, where there is no chilliness. Imagine the labor — each one of those thousands of larvæ carried separately in the mouth of a faithful nurse! If a shower comes on, or if the young family is threatened with danger, they are quickly taken to some safe place.

7. When ready to enter the pupa state, the larvæ cover themselves with a sort of web (Fig. 86), and are still carried back and forth by the nurses, who continually clean them. Sir John Lubbock, in his

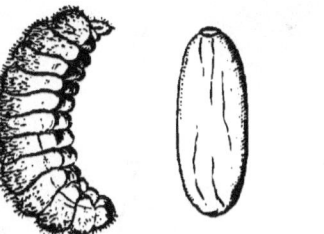

Fig. 86.—LARVA, COCOON, AND PUPÆ OF RED ANT (MAGNIFIED).

recent work on ants, states that when the pupæ are ready to leave their cases the nurses help them to escape. "It is very pretty," he says, "to see the older ants helping them to extricate themselves, carefully unfolding their legs and smoothing out the wings with truly feminine tenderness and delicacy."

8. **Cleanliness.**—Ants not only keep their homes neat, but they are careful of their own personal cleanliness. Their little feet are covered with hairs, which form good brushes, and no particles of dust are allowed to remain on their bodies. They may often be seen rubbing their feet together to clean them, as flies do. The antennæ of ants (Fig. 87) are bent like an elbow, and with them the active little creatures examine every object they meet.

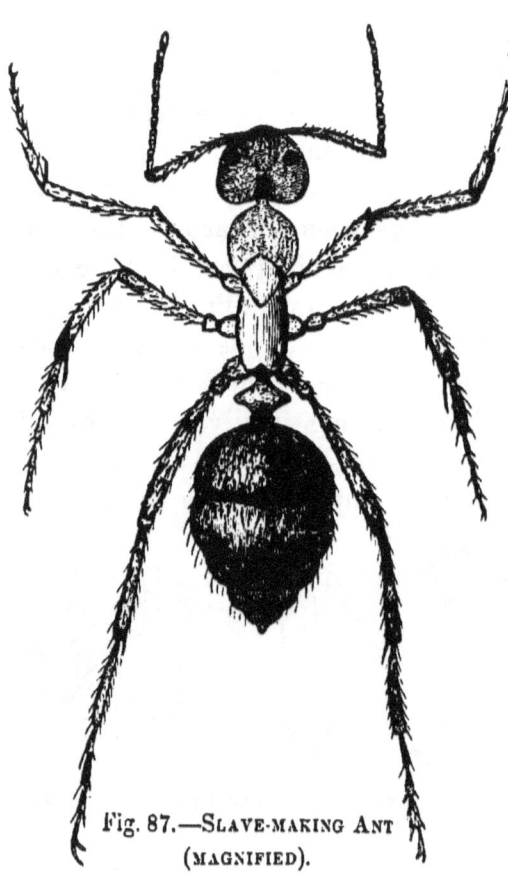

Fig. 87.—SLAVE-MAKING ANT (MAGNIFIED).

9. **Recognizing Members of their own Family.**—If we notice ants in their travels, we shall see two distinct lines, one moving towards the nest, the other leaving it. Those that are returning without a load stop, and, with their antennæ, salute their sisters carrying burdens, and this they do so quickly as not to break or interrupt the line.

10. In one nest there may, perhaps, be four hundred thousand ants. Notwithstanding these immense numbers,

a stranger upon entering the nest is immediately attacked, which fact shows that the ants in the community have some power of recognizing each other. They even know members of their own family after a long absence, and welcome them back to their old home.

11. If an ant has discovered a good feeding-ground, it seems to spread the news to its fellows, and often returns with a troop of them to share the feast.

12. **Favorite Places for their Nests.**—You have probably noticed the little ants that burrow under the pavements in our streets and door-yards, and have wondered why they choose situations so exposed that many of them are trodden underfoot, while their little hillocks of earth are swept away by the broom.

13. We may rest assured that they have good reasons for this singular choice, and that the situations are not undesirable, or the ants would not seek them.

14. In the first place, the ants must have a care to supply their growing family with food, and where could they fare better than near the homes of man? The tiny crumbs dropped by the children are treasures to the economical ants, whose sharp eyes see many chances for feasting upon things we have thrown aside as useless.

15. Then, too, the bed of fine gravel which the brick-layer smooths so carefully to lay his bricks on is a fine place for the ants to burrow in. The sun, shining upon the bricks, heats them, and also the earth beneath, and makes a warm place for the ants to put their larvæ when they bring them up out of their nests.

16. You know how common it is, on turning over large stones, to find the ground beneath covered with the white larvæ of ants, which are quickly carried away and hidden. The stones become heated during the day, and retain the

heat long after the sun has set. Ants, no doubt, select these spots that they may secure a safe, warm place in which to hasten the development of their larvæ and pupæ.

17. **Singular Relations with Plant-lice.**—Ants feed chiefly upon insects, killing great numbers of them, and they also eat honey, fruit, and almost any sweet substance. This liking for sweets has led them to form singular relations with our common green plant-lice, the aphides. The plant-lice secrete a sweet liquid called honey-dew, of which ants are very fond, and which they obtain by tapping the lice with their antennæ.

18. Some species of ants ascend into bushes in search of these lice, and, having found them, watch over and defend them from attacks by other insects. Sir John Lubbock says that the ants take care of the brown eggs of aphides during winter, carrying them to the lower chambers of the nest when it is disturbed. In the spring, when the young aphides hatch, they are brought out and placed on tender shoots of plants.

19. This is a remarkable instance of forethought. The ants derive no immediate benefit from the eggs, yet by taking care of them they secure a supply of their favorite honey-dew for the following summer.

20. **Capturing Slaves.**—Fierce battles are fought between different colonies of ants apparently for the sole purpose of capturing slaves. This instinct is so strong with the common red-ant that it is spoken of as the "slave-making ant." It frequently invades the nests of black-ants, and fearful struggles occur between the two colonies.

21. When about to attack the enemy, red-ants leave the nest in full force and march directly to the battle-field. It is not a general warfare, but each red-ant seizes upon some black one, and makes a desperate effort to kill it.

After the battle, if the red-ants are victorious, they enter the conquered nest and carry off the larvæ and pupæ, which they bring up as slaves. These young slaves enter at once upon a life of toil, and make no effort to escape.

22. **Degrading Effects of Slavery.**—It has been noticed that this system of slavery has a degrading tendency among ants, as it is well known to have among men. Some of the slave-making ants are so accustomed to being waited upon that they have lost the art of building and of caring for their young, and are entirely dependent upon their slaves for these services.

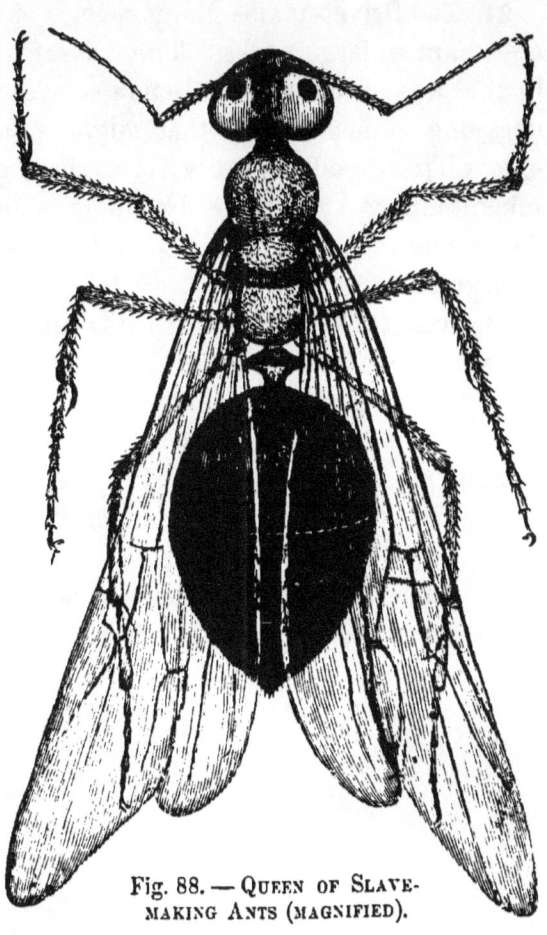

Fig. 88. — QUEEN OF SLAVE-MAKING ANTS (MAGNIFIED).

They have even lost the habit of feeding themselves, and, although surrounded by food, they will starve unless fed by others.

23. **The Harvesting Ants.**—The harvesting ants of Texas

clear a circular space, ten or fifteen feet in diameter, around the entrance to their nests. Within this space nothing is allowed to grow but "ant rice"—a species of grass, the seeds of which are carefully gathered by the ants.

24. **The Driver-ants.**—Many species of ants in hot countries hunt in large packs. The driver-ants of Africa hunt in this way, and render valuable service in clearing away decaying animal matter that might otherwise cause disease. The dread of visits from these ants compels the inhabitants to keep their dwellings comparatively clean. These hunting ants are said to be blind, and go out chiefly at night.

25. **Termites.**—Termites, or white-ants, as they are called, do not properly belong here, as they are not true ants. Still, we will study something about them. They abound in all tropical countries, living in large communities and committing serious ravages. They build structures aboveground, often five feet high, composed of earth worked and patted until it becomes nearly as hard as stone. There is no external opening in these hills, but the entrance is placed at some distance, and is reached by underground galleries.

XXII.

OYSTERS.

SUB-KINGDOM, MOLLUSCA : CLASS, LAMELLIBRANCHIATA.

1. **Mollusks: the Mantle.**—Having taken this mere glance at the Articulates, we will begin our study of Mollusks. This division includes soft-bodied animals which are usually provided with shells, and which pass by the general name of "shell-fishes." Their bodies are enclosed by a delicate membrane called a "mantle," whose office it is to secrete the shell. On opening an oyster we see this thin, glistening mantle lining the shell as well as covering the oyster.

2. **The Shell adapted to its Surroundings.**—The shell is useful in protecting the soft body of the mollusk, and its strength and thickness are generally in proportion to the dangers to which the animal is exposed. Those species inhabiting shallow places in the ocean near the shore, and hence exposed to the beating of the waves, have stronger shells than those living in deep water. Fresh-water mollusks, on the other hand, generally have delicate shells.

3. Another provision of nature for the safety of the helpless mollusks may be seen in their coloring. Those which spend most of their lives at rest near the same spot, as oysters and clams do, are of the same general color as their surroundings. On the contrary, those that move about, as pectens and gasteropods, are often tinted with rich and beautiful colors.

4. A Bivalve Shell.—When a shell consists of two separate pieces or valves opening by a hinge, it is called a bivalve.

5. The Shell of an Oyster examined.—A careful examination of one mollusk will help us to understand all the others ; therefore we will take an oyster as our type. If we cannot obtain a living oyster, let us at least have the shell, and examine it carefully. What is the first thing you see? Is it the thin layers of which the shell is composed?

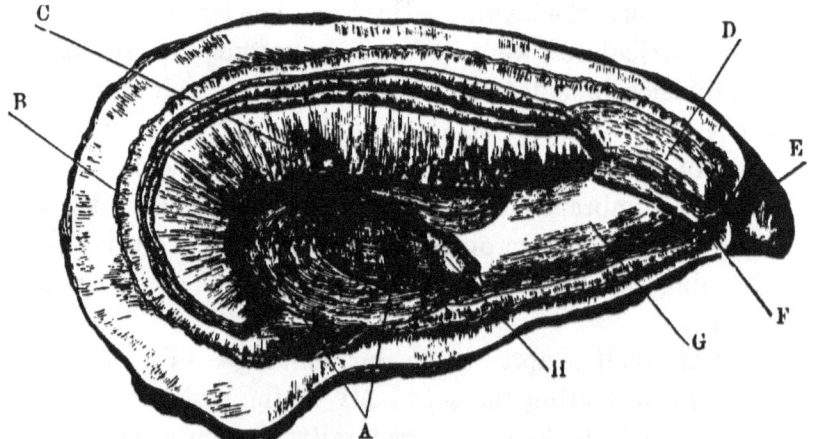

Fig. 89.—Oyster in the Shell (with Mantle removed from the Upper Surface).

A, muscle; B, mantle; C, gills; D, labial palpi; E, hinge; F, mouth; G, liver and stomach; II, heart.

6. These layers are very interesting. You will soon suspect that they have been caused by the growth of the oyster. By looking on the outside of the shell you may see the lines of growth, and perhaps you can detect the shape of the oyster when it was very small. The delicate mantle (B) has deposited new layers of shelly matter upon

the inside from time to time, each layer extending a little beyond the edge of the last, and increasing the size of the shell.

7. After an oyster has obtained its full growth the shell does not increase further in size, but it becomes thicker by the addition of one layer inside of another, so that the age of an oyster may be estimated by the thickness of its shell. This thickening is readily seen at the hinge (E), which seems to have grown in until it encroaches upon the space intended for the oyster. Yet you will see that at one time the hinge was at the very tip of the beak.

8. In a freshly opened oyster you will notice a tough brown band in the hinge; this is the ligament which unites the two valves, but, strangely enough, it acts like a spring which is constantly tending to throw the shell open. Let us see what causes this. The elastic, horny fibres which form the ligament are placed endwise between the valves; consequently, they are squeezed when the shell is closed, and they try to make room for themselves. If the ligaments in the hinge push the shell open, how, then, do you suppose it can be closed?

9. The purple spot on the inside of each valve shows where a muscle was attached which extends right through the body of the oyster (A), and holds the two valves together. You know the oysterman has to cut the oyster loose from the shell at this point with his knife, and this is the only place at which the oyster is attached to the shell. The muscle is the tough part of the oyster, and when it shortens itself the valves are drawn together. If the muscle lengthens, the valves fly open, as is the case when the oyster dies. Bivalves naturally stand open with a stream of water flowing over the gills, unless they are forcibly held together by the muscle. Fortunately for us,

7

oysters live some time after being taken out of the water, but they keep their valves closed to retain their moisture.

10. The inside of the shell is further marked by the "pallial line," which shows where the edge of the mantle has rested. By lifting the fringed edge of the mantle four delicate gills (C, Fig. 89) may be seen extending part way round the edge of the oyster. The gills are covered with cilia, which by rapid motion produce a current of water towards the mouth, bringing to it particles of food and, as the current flows away again, it carries off the waste matter.

11. **Food floated to the helpless Oyster.** — The helpless oyster, fastened down to its bed, has no possible way of seeking food, and it is therefore entirely dependent upon these currents of water. Coming in this way, the food necessarily consists of very small plants and animals, which are abundant in the sea, especially in the quiet places where oysters flourish. Oysters live in shallow water attached to some fixed object by the lower valve, which is larger and deeper than the other; in it the oyster lies as in a trough.

12. **The Oyster examined.**—The mouth is a mere slit at the smaller end of the oyster (F), near the hinge. It is covered by four thin lips or folds of membrane, called "labial palpi" (D). An œsophagus leads to the stomach, and the intestine passes through the large liver (G), which is of a brownish-green color, and occupies most of the soft part of the oyster.

13. Oysters have no true head. The heart (H) may easily be seen in a clear space near the muscle, and in a freshly opened specimen it will beat slowly and regularly. It consists of two sacs, one large and transparent, the other small and brownish.

14. **Large Number of Young Ones.**—Spawning season occurs during the summer months, at which time the eggs floating in the fluid around the gills give to it a thick, creamy appearance. Oysters are not then in good condition for food. They produce an immense number of young ones. It is thought one oyster may yield a million in a season, and the whole number of young oysters thrown out from an ordinary oyster-bank is almost incalculable. The eggs leave the parent shell in puffs of milky fluid, and are sometimes so thick as to make the water look clouded until they are scattered by the waves.

15. **They settle for Life.** — When the young ones are hatched they swim about for a time, then attach themselves for life to some solid object. Unless they find a clean, hard surface to fasten to, the little things will probably perish in the mud or be devoured by larger animals. A very large proportion of the young is destroyed in this way.

16. **Oyster-beds.**—Oyster-beds generally exist in brackish water upon a bottom of clay or mud which is firm enough to prevent the oysters from sinking into it. The water must also contain lime to supply the oyster with the material for its shell. It is found that oyster-beds increase in the direction of the current, the young ones having drifted with the tide before settling.

17. **Cultivation of Oysters.**—In addition to natural oyster-beds, there are many "oyster farms," where these delicious mollusks are regularly cultivated. Stakes are driven in the mud in shallow water, and branches of trees, rough boards, or stones are placed between them for the baby oysters to fasten themselves to. When the nursery is ready, several boat-loads of oysters are dropped near the spot. The oysters rapidly increase in size and num-

bers, and are ready for the table in from two to four years.

18. Oysters destroyed by Dredging.—Oysters are generally fished with a dredge. As this instrument is dragged over the bed, the teeth pull up the oysters, both large and small, from their resting-place. Those that are too young for market are thrown back into the water, and if they fall on a suitable surface they will again attach themselves, and continue to grow. Many of them, however, sink in the mud and are suffocated.

19. The process of dredging is also destructive to the oysters which remain on the bed, as they are roughly torn from each other and dragged into the mud. Here they cannot open their valves without admitting the mud, and this is certain death to an oyster.

20. Oysters are highly esteemed for food on account of their delicious flavor, and the demand for them is constantly increasing. This leads to excessive fishing of the oyster - beds, and in many places the beds yield a much smaller supply than formerly. Such is the case with many of the European oyster - beds. The French government has been obliged to take control of those on its shores, and to enforce certain laws with regard to fishing them.

XXIII.

MUSSELS AND PECTENS.

SUB-KINGDOM, MOLLUSCA : CLASS, LAMELLIBRANCHIATA.

1. **Marine Mussels.**—Marine mussels grow in large beds in shallow water, fastened to stones and sand-banks, and making a solid black mass. They often cling to posts and piers, where they are left uncovered when the tide is low. At such times they keep their shells tightly closed, like barnacles.

Fig. 89a.—BUNCH OF MUSSEL-SHELLS.

2. **The Foot of the Mussel.**—The structure of mussels is similar to that of oysters, except that they have a tough foot. This is a thick, fleshy organ, which may be pushed out to a great length. In different species of mollusks the foot has various uses, enabling the animal to push it-

self about or to leap, while often it is used for boring holes in the sand or mud. Although this organ helps some mollusks to move about from place to place, it does not resemble a real foot, but is more like a tongue.

3. **The Byssus.**—Mussels are hatched within the shell of their parents. After leaving the shell, and swimming around for a while, they attach themselves to some object by silken threads called byssus. At the base of the foot is a gland for secreting the fluid byssus, which, when dry, forms into brown threads not unlike the silk of spiders and caterpillars. The foot attaches this sticky fluid to some object, and is then withdrawn, leaving the silk fastened to the surface. Mussels are also joined to one another in great bunches, as well as to the bed of the ocean.

4. The threads of byssus are long enough to admit of slight motion, as the mussels float and drift back and forth, so these animals are not compelled to remain in one position, as oysters do. If the byssus is broken the mussels attach themselves again by other threads.

5. How strange it seems that these lowly sea-creatures should spin silk ; yet the long, fine threads of byssus have sometimes been woven into gloves and stockings, and even into cloth.

6. The fresh-water mussels have no fierce waves and tides to resist, and therefore do not secrete byssus.

7. **Mussels cultivated for Food.**—Salt-water mussels are used for food, and are cultivated like oysters. When the young mussels have reached the size of a small bean they are scraped in masses from objects to which they have adhered, and are carried in baskets to places suited for their growth. They soon attach themselves to posts and branches of trees prepared for them, and are transplanted in this way three times before reaching their full size.

8. **The Epidermis of Shells.**—All living shells have an outer layer of animal matter called epidermis ; they have no lustre upon the exterior until this epidermis is taken off and the surface is polished. Mussel-shells show beautiful blue tints when the epidermis is removed.

9. **The Color heightened by the Action of Light.**—The color of shells depends much upon the action of light, and those grown in shallow water have generally brighter colors than those grown in deep water. The largest and most highly colored shells are found in the tropics, whereas arctic shells are mainly small and dull. The peculiar lustre of shells is due to the minute edges of alternate layers of carbonate of lime and animal tissue. In order to fully enjoy these treasures of the ocean, we must see them under the sparkling water, where their beautiful forms and colors are heightened by

"The sun, and the sand, and the wild uproar."

10. **How Pearls are Formed.**—Pearls are formed in shells when grains of sand lodge between the mantle and the shell and become coated with the shelly matter, or "nacre," which the mantle secretes. Fresh - water mussels yield pearls that are sometimes quite valuable, but the finest pearls are obtained from the pearl-oyster. The pearl-oyster in Fig. 90 is the circular shell, which has a straight hinge and one pearl clinging to it, and which is partly covered by the mussel-shell.

11. Pearls mostly have a nucleus of sand in the centre, and the shelly layers are arranged around it like the coats of an onion. The Chinese take advantage of this fact, and sometimes place small images or beads inside the shell, allowing them to remain until they are coated with pearl. Some of these are shown at the right of the picture.

Fig. 90.—PEARL-BEARING SHELLS.

12. **Pearl Fisheries.**—The most important pearl fisheries are on the coast of Ceylon. The same locality is not fished every year for fear of impoverishing it. The labor of diving for pearl-oysters is very severe. The divers remain under water only thirty seconds at a time, but they sometimes dive twenty times in one morning, and become very much exhausted. Having touched bottom, the diver gathers the oysters within reach, and places them in a net, then he pulls a cord as a signal to be drawn up immediately. At mid-day a gun sounds for the fishing to stop, and the boats are taken to the shore and unloaded before dark, in the presence of officers, so that no robbing shall be done.

13. The oysters are allowed to remain on shore until they decompose. The pearls are then easily gathered from the gaping shells, and they are worked with powdered nacre to give them a good polish. Pearls may be round, ovoid, or pear-shaped. Those which adhere to the

valves are consequently irregular in shape, and, as they are not so valuable as others, they are sold by weight. Mother-of-pearl is the lustrous layer taken from the inside of the shell of the pearl-oyster.

14. **Celebrated Pearls.**—There are a few fine pearls so remarkable for their size and beauty as to have become historical, and their line of descent can be traced for generations. Most of these pearls belong to kings and princes. A famous string of pearls belongs to the Shahs of Persia in which each pearl is the size of a hazel-nut.

15. **Pectens.**—Pectens (Fig. 91) are found in all seas, and of many different varieties. Their elegant shells are ribbed and mottled with various colors, and they grow by additions made to the edge, rather than by a thickening of the valves, as in the oyster. The hinge is extended into broad ears, and is worked by a ligament placed in a tiny pit which you can easily discover. The mantle is fringed with tentacles, and has a double row of bright spots on the edge, which are thought to be eyes.

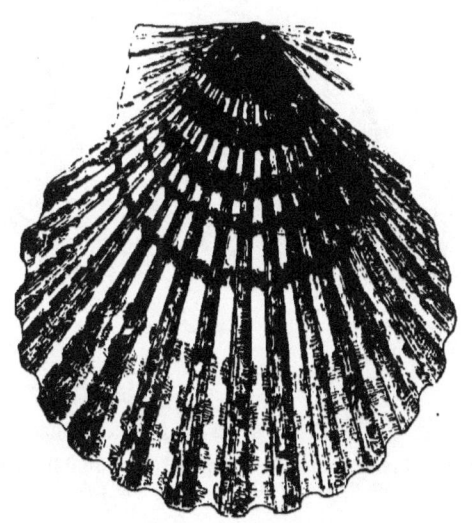

Fig. 91.—Pecten Shell.

16. Contrary to the habits of most bivalves, the pretty little pectens can swim through the water. As they are propelled by alternately opening and closing their valves, their movements consist of a succession of jerks.

7*

XXIV.

CLAMS AND RAZOR-FISHES.

SUB-KINGDOM, MOLLUSCA : CLASS, LAMELLIBRANCHIATA.

1. **Markings upon the Inside of a Clam-shell.**—The name "clam" is applied to many different species of mollusks along our coast having thick shells. Upon looking care-

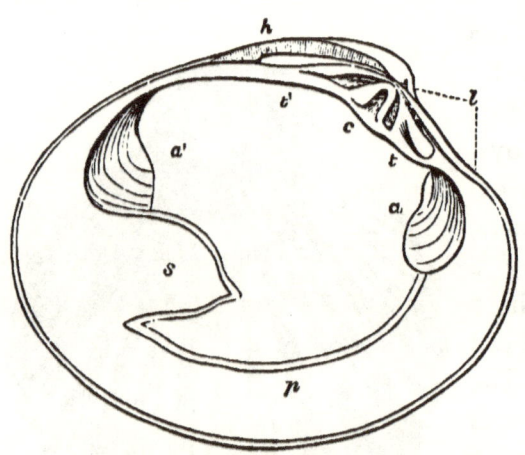

Fig. 92.—INSIDE OF A CLAM-SHELL.

a, a', impressions of the muscles; p, pallial line; s, bend occupied by the siphon; h, hinge; c, t, t', teeth.

fully at one of these shells you will find that it differs in many respects from the oyster-shell which we examined. You will at once notice the two marks (a, a', Fig. 92) left by the muscles, and you will readily infer that the valves of clam-shells must be connected by two muscles.

2. What curious freaks the pallial line (p) has taken between these two points. You can scarcely understand now why it should turn inward and make that deep bend, but when you have learned about the animal that inhabited this shell, the reason will be plain to you.

3. **Teeth in the Hinge.**—There are also peculiarities in the hinge (h) which we must not fail to observe. The spaces between the three teeth (c, t, t') are exactly fitted by two other teeth in the opposite valve, and these interlock when the shell is closed. In some species of clams there is a large, spoon-shaped hollow at the hinge, with long ridges on each side fitting into corresponding grooves on the opposite valve. The central hollow space contains the ligament, or spring, which, as we learned in the oyster, is always trying to push open the shell.

4. **Lines of Growth on the Shell.**—On the outside of the shell the lines of growth are plainly seen, and you can trace the exact size of the clam at different periods of its history all the way back to babyhood. These shells do not grow thick with age.

5. **Mantle attached to both Valves.**—A clam, we know, always looks torn and ragged on opening the shell. It is impossible to remove the valves and leave the animal smooth and uninjured, as the oyster is when taken from its shell. This is because the mantle is attached to both valves along the pallial line, making a closed bag for fluids, which is torn when we open the shell.

6. **Mantle rolled into Tubes forming a Siphon.**—Water is admitted into this closed sac only through a siphon (b, c, Fig. 93), which is in reality the mantle rolled up into two tubes. Through one of these tubes a stream of sea-water enters, and, circulating under the mantle, passes down to the mouth and gills. It is then thrown out by the second tube, carrying off with it all waste matter. The circulation of water is kept up by countless cilia which line the tubes, and which, by their constant waving motion, draw the water towards the gills.

7. The tentacles at the entrance of the siphon are very

sensitive to the touch, and keep out all floating particles except the very small ones which are suitable for food.

8. You will now understand that the curious bend (*s*) in the pallial line (Fig. 92) is the impression left by this siphon.

9. **Organs.**—The gills and the labial palpi of the clam are similar to those of the oyster. The heart is under the hinge, and, strangely enough, the intestine passes directly

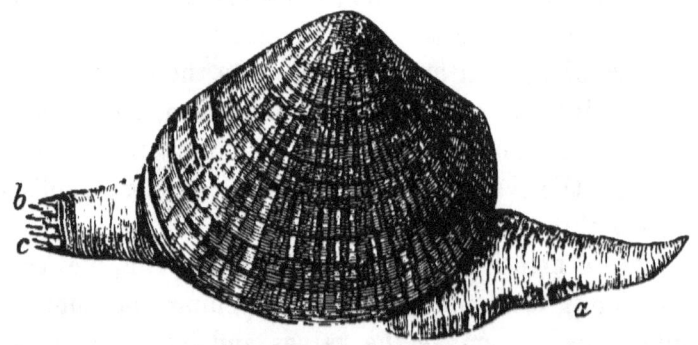

Fig. 93.—CLAM (MACTRA).

a, foot; b, c, siphons.

through it. Clams have a tough foot (*a*, Fig. 93) near the mouth, by means of which they push themselves along and dig holes in the sandy beaches, and to this life they are well suited. It is surprising to see how these animals can increase their size when they wish to extend the foot, the siphon, or the edges of the mantle. This is done by taking in sea-water through numerous pores in the skin. Touch the mollusk, however, when these parts are extended, and they are quickly drawn in and the shell closed.

10. **Clams lie Buried in the Mud.**—Clams spend their time buried in the soft mud, with the mouth downward and the siphon extended far enough out of the shell to reach the water above. They may sometimes be seen

spouting water from small holes on the beach.   It is good sport to dig them out and see how nimbly they bury themselves again in the sand, using no tool but the foot.   Many clams have only a short siphon which does not extend far beyond the shell.

11. **Razor-fishes.** — Some of these holes on the beach you may find occupied by razor-fishes (Fig. 94), which are not so easily caught as clams.   These mollusks are abundant on all sandy shores, where they live buried in the mud.   By means of the foot they dig a deep hole, which they do not leave.   They raise themselves to the entrance of this hole, but disappear quickly upon the slightest alarm.

12. Fishermen become very expert in dealing with the peculiar habits of timid sea-animals, but even the fishermen find the razor-fish hard to catch, and if they fail in the first attempt to capture it, no further efforts will induce the shy creature to appear again.

Fig. 94.
RAZOR-SHELL
(SOLEN).

13. The long, slender razor-shell is thin and brittle, with delicate tints of rose or violet, which are nearly concealed by the brown epidermis covering it.

XXV.

GASTEROPODS.

SUB-KINGDOM, MOLLUSCA : CLASS, GASTEROPODA.

1. **Gasteropods a large Class.**—Leaving the bivalves, we will now turn our attention to the gasteropods—a large class, which contains three-fourths of all living mollusks. They are found in fresh water as well as in salt. Some, such as snails, live also on the land.

2. **Univalves.**—Gasteropods are known as univalves, since they have but one shell, which is generally a tube twisted spirally from a point called the apex. A few of their beautiful forms are given in Fig. 99. In almost any collection of shells you will find some of these gasteropods. By sawing one open the spiral tube may be seen winding round a central column, as shown in Fig. 95, and gradually growing larger towards the opening. You will be interested in tracing the coil on the outside of these shells, observing that as the occupant increased in size it made for itself more and more room in the shell.

Fig. 95.—SECTION OF A SPIRAL UNIVALVE.

3. **Shell enlarged by Secretion from the Mantle.** — We have before learned that all shells are secreted by the mantle. As the shell of the gasteropod needs enlarging, the mantle, stretching over the edge of it, deposits a layer of shelly matter, and thus the shell gains a new and larger rim. The outer edge of the mantle often contains bright-colored spots, which impart their tints to the rim of the shell, ornamenting it with bright streaks and lines.

Fig. 96.—WHELK.
o, operculum; s, siphon.

4. **What Forms the Spines on some Shells?**—The edges and notches of the old rim are often marked upon the outside of the shell, and there are sometimes long, bristling spines sticking out from them. How could these spines have been formed? Wherever there is a spine, there must have been at that point a fold of the mantle pushed out over the rim of the shell to form a tube. This fold, like every other part of the mantle, deposited shelly matter, and finally formed the stiff spine. Of

course, it is of no further use after the rim has grown beyond it.

5. Most of those gasteropods that have the margin of the shell notched and lengthened into a canal are flesh-eaters, whereas those having an entire and even margin live on vegetable food.

6. **Gasteropods Highly Organized.** — Gasteropods, as a general thing, are quite highly organized. They have a distinct head, with two tentacles, and eyes that are some-times stalked; they are believed to have the senses of hearing and of taste, also, which indicates a high-er stage of devel-opment than that of the oyster and clam.

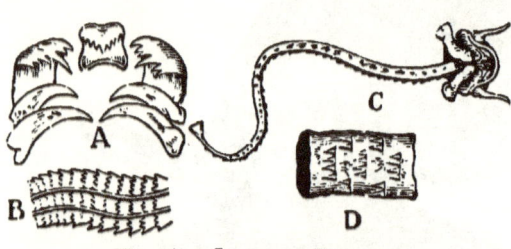

Fig. 97.—LINGUAL RIBBONS.

A, portion of tongue of *Velutina*, enlarged; B, portion of tongue of whelk, magnified; C, head and tongue of limpet; D, portion of same, greatly magnified.

7. **Siphon— Operculum.** — Water is admit-ted within the body by means of a siphon, and at this point the shell is often lengthened into a long canal. The thick tough foot may be extended entirely beyond the shell, but gasteropods are timid creatures, and when alarmed all parts of the body are instantly drawn in, and the entrance is closed with a horny plate on the foot, which is repre-sented at *o*, Fig. 96. This plate fits snugly in the shell, and is called the operculum. The operculum of some gasteropods consists of limestone; small ones of this kind are known as "eye-stones," and were formerly used to re-move irritating particles of dust from the eye.

8. **Lingual Ribbon.** — Gasteropods have a remarkable tongue, which contains many sharp-pointed teeth set in

distinct rows (Fig. 97). The growth of the tongue continues during the life of the animal, new teeth forming at the base of the tongue and growing forward to take the place of those that are worn off at the tip. This tongue is spoken of as the "lingual ribbon," or as the "odontophore."

9. Shells bored by the Lingual Ribbon.—With the lingual ribbon gasteropods file holes in other shells, through which they suck out the soft body. Many strong shells that would apparently make an excellent defence are found to be pierced in this way by a round hole, the

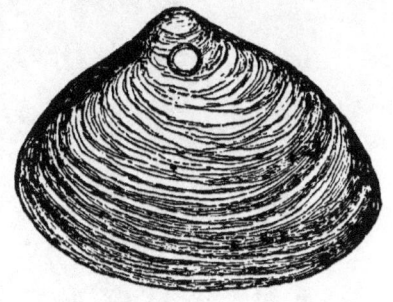

Fig. 98.—CLAM-SHELL BORED BY LIN-GUAL RIBBON OF GASTEROPOD.

edges of which are perfectly true and even, indicating not only good tools, but a skilful use of them (Fig. 98).

10. Shells that are washed ashore are mostly empty, and now that your attention has been called to the fact, you will be surprised to see so many bearing this round hole, and telling the sad fate of their former inmates. You will find that the hole is made near the hinge, and directly over the softest part of the body.

11. **Digestion of Gasteropods.**—In addition to the numerous teeth on the tongue of gasteropods there are hard plates in the stomach for crushing food. After being mixed with saliva, which is furnished by salivary glands, the food passes through a long œsophagus into the stomach. Here the food is acted upon by fluids secreted by the liver and other glands. It then passes into a long intestine, where the nourishing portions are absorbed into

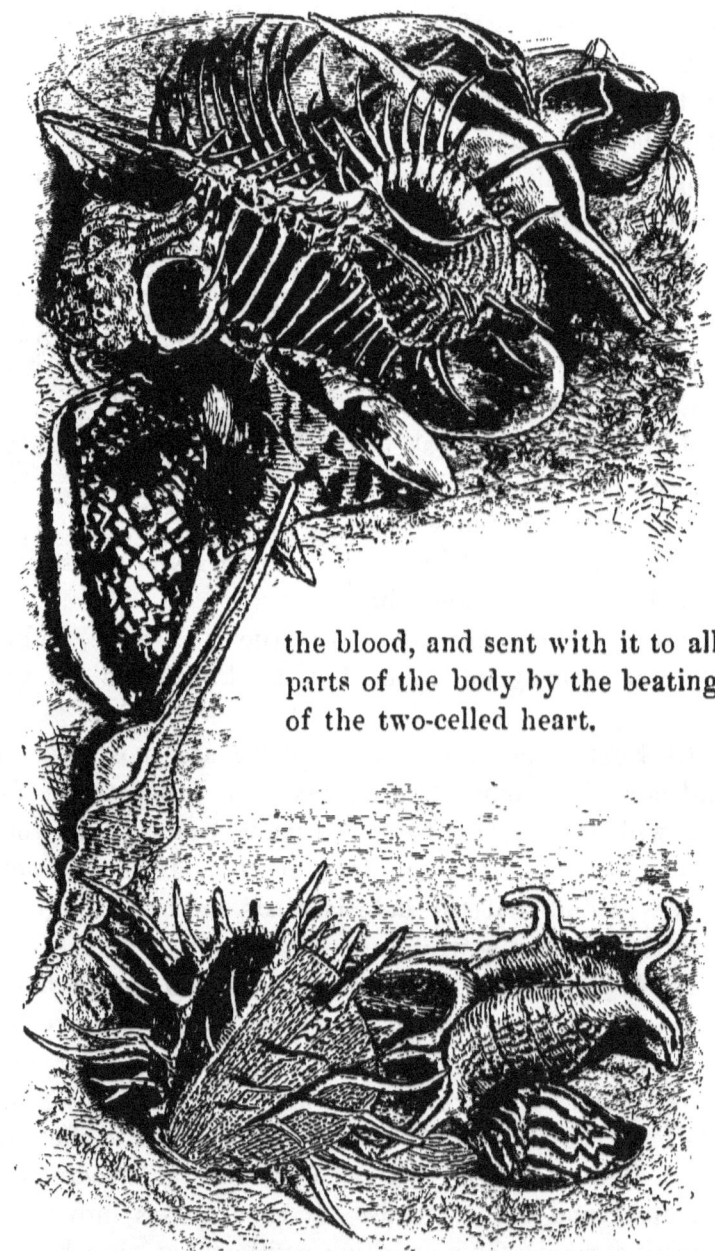

the blood, and sent with it to all parts of the body by the beating of the two-celled heart.

Fig. 99.—SEA-SHELLS.

12. **Breathing.**— Gasteropods breathe either by lungs or by gills, some of them coming frequently to the surface of the water for air. They push themselves along by the foot, and many of them swim freely through the water.

13. **Clusters of Egg-cases.**— On the sea-shore we find many singular-looking objects, whose appearance alone would give us no idea of their real character. This is true of the odd-shaped egg-cases of many gasteropods.

Fig. 100.—EGG-CASES OF WHELK.

Among these are the eggs of the whelk (Fig. 100), which are found united in large masses, each egg in the cluster being enclosed in a little sac of its own.

14. Many of you have picked up on the sea-shore long strings of the egg-cases of the pear-conch (Fig. 101). These are very common on sandy beaches. They are composed of many cream-colored cases, or capsules, of a tough, leathery substance, which diminish in size towards both ends of the string.

15. They contain eggs which hatch within the capsule, each little conch being provided with a tiny shell. After consuming the jelly-like fluid with which the capsule

Fig. 101. — EGG - CASES OF PEAR-CONCH.

is filled, the animals work their way out of the leathery bag and bury themselves in the sand.

16. If you examine the cases you will find a little round hole on the top of each one, which is closed by a

gristly substance, and looks as if it were provided as an easy means of escape for the young conchs. Cut open a case, and if the little occupants have not already escaped, you will find it filled with lovely shells.

Fig. 102.—NATICA.

17. **The Nidus of the Natica.**—One of the sea-snails (the natica, Fig. 102) makes a ring-shaped nest, or "nidus," of fine grains of black and white sea-shore sand, glued together by the slimy substance in which the eggs are deposited. This nidus (Fig. 103) when first taken from the water is soft and leathery, but it becomes exceedingly brittle when dry. It somewhat resembles the broad rim of an old felt hat, and its surface is often thickly studded with the

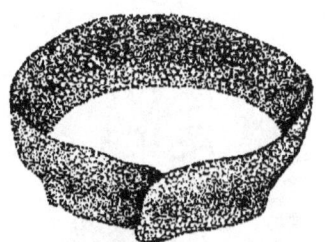

Fig. 103.—EGG-CASE OF NATICA.

egg-capsules of another gasteropod, the *nassa*, which avails itself of this convenient place of deposit for its eggs.

XXVI.

LIMPETS AND LAND SNAILS.

SUB-KINGDOM, MOLLUSCA : CLASS, GASTEROPODA.

1. **Limpets.**—Limpets are attractive little gasteropods living on those parts of the sea-shore which are left un-covered at low tide. Our best time to watch them will be when the shallow water is rippling over their bodies, and their conical shells are lifted that they may enjoy the full benefit of the bath.

2. **The Shell.**—The univalve shell, you will observe, is not spiral, but is a simple oval shell, tapering to a point on the top like a tent. This shape gives great strength to the shell, and enables it to support a heavy weight without injury. The exterior of the shell is a dull gray color, without much ornamentation, but the interior is peculiarly smooth and lustrous, and is prettily marked by the pallial line.

Fig. 104.—LIMPET-SHELL.

3. When under water, limpets move about slowly by means of a round foot, but as their gills cannot long bear exposure to the air, when the tide is out their shells are drawn down close to the rock and held there tightly.

4. **Limpets adhere firmly to the Rocks.**—The foot has some power of adhering firmly to surfaces as if it were glued, and when the limpets are alarmed they hold on so

tightly that it is difficult to remove them. In attempting to pry them off, the shell is often broken before they let go their hold. Limpets sometimes remain so long in one spot that the rock is hollowed out to their exact shape. Sea-birds are fond of eating them, and are so cunning as to thrust their bills under the shell when it is lifted.

5. **How Limpets Eat.**—The limpet's head is furnished with a pair of eyes and a pair of tentacles. The lingual ribbon is covered with sharp teeth set in three rows, and is three times the length of the entire animal. Limpets feed upon sea-weed, sometimes making a noise with the tongue as it strikes upon the hard upper jaw in biting.

6. **A Large Limpet.**—The limpets in tropical seas are larger and richer in color than ours. One species is found at the Strait of Magellan having a shell nearly a foot in width, which is used by the natives as a basin.

7. **Snails.**—Limpets furnish us an example of gasteropods that live partly out of water, but their cousins, the snails, which we will next consider, have gone a step farther and live altogether on land.

8. It is amusing to watch the motions of these curious snails as they crawl about with their great shell houses on their backs, stretching out their feelers, then suddenly drawing them in again. All at once some fancy seems to strike these uncertain individuals, and the whole slimy dark-gray body is pulled back into the shell.

9. **The Shell.**—The shell is remarkably light and delicate, and you may easily trace the coil upon the outside. In some species the edge is plain and sharp, while others have the edge folded back to make a smooth, firm border.

10. **Senses of Sight and Smell.**—Snails are better travellers than limpets, and far more active. Like them, they have a foot and a lingual ribbon. Besides the long tenta-

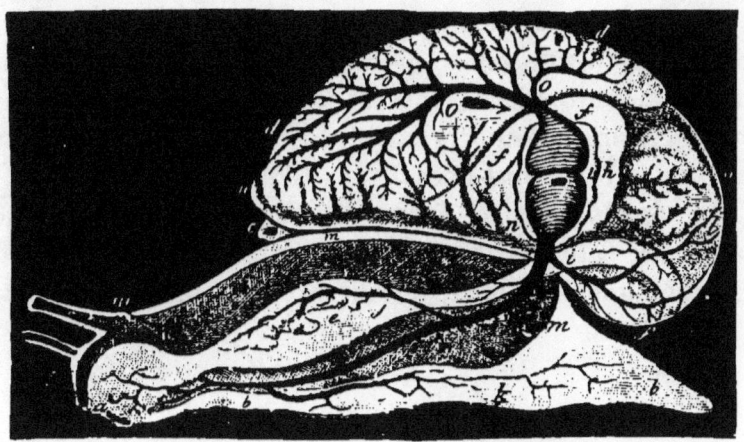

Fig. 105.—Anatomy of a Snail.

a, mouth; *b*, foot; *c*, anus; *d*, lung; *e*, stomach, covered above by the salivary glands; *f*, intestine; *g*, liver; *h*, heart; *i*, aorta; *j*, gastric artery; *k*, artery of the foot; *l*, hepatic artery; *m*, abdominal cavity; *n*, irregular canal communicating with the abdominal cavity, and carrying the blood to the lung; *o*, vessel carrying the blood from the lung to the heart.

cles tipped with black eye-specks, snails have a shorter pair, which, it is thought, are organs of smell. This latter sense is apparently more keen than their sight, since they are attracted by odors of fruit and vegetables, though they do not seem to see obstacles placed in their way.

11. **The Breathing Organ.**—The breathing organ of snails is a chamber lined with a net-work of blood-vessels (*d*, Fig. 105), and supplied with air by a small orifice which may be seen to open occasionally. The air is then expelled from this chamber by drawing the body into the narrow part of the shell.

12. **Where Snails Live.**—Snails delight in warm, damp weather, and they may be easily found in shady places in the woods. When winter comes they hide in the ground, and close their shells with successive layers of mucus, which, when dry, form a hard membrane over the opening. Their eggs are laid loose under damp leaves and stones.

13. These land mollusks have perhaps gradually accustomed themselves to living, first in marshes, then in damp, swampy places, until finally we have some species living entirely upon dry land. Still, their favorite spots are the shady, moist ones.

14. **Snails eaten for Food.** — In many parts of Europe snails are eaten for food, and they are sometimes painted on the sign-boards of restaurants and drinking shops.

Fig. 106.—An Edible Snail.

They were considered a delicacy by the ancient Romans, who served them at their funeral entertainments. In the buried city of Pompeii, among other curious relics, heaps of snail-shells, which are the remains of these funeral feasts, are found in the cemetery.

15. **Young Snails.** — Common snails kept through the winter in the damp earth of our window plants will prove a source of great interest. In the spring they deposit tiny white eggs, so delicate as not easily to be lifted. On breaking one, a perfect little snail-shell may be seen within. Later on we have the gratification of seeing the young snails start off for themselves, creeping up

and down over the rough places, and performing, on a small scale, all the manœuvres of their elders.

THE HOUSE-KEEPER.

"The frugal snail, with forecast of repose,
Carries his house with him where'er he goes;
Peeps out, and if there comes a shower of rain,
Retreats to his small domicile amain.
Touch but a tip of him, a horn, 'tis well—
He curls up in his sanctuary shell.
He's his own landlord, his own tenant; stay
Long as he will, he dreads no quarter-day;
Himself he boards and lodges; both invites
And feasts himself; sleeps with himself o' nights.
He spares the upholsterer trouble to procure
Chattels; himself is his own furniture,
And his sole riches. Wheresoe'er he roam,
Knock when you will, he's sure to be at home."

CHARLES LAMB.

8

XXVII.

THE OCTOPUS, OR DEVIL-FISH.

SUB - KINGDOM, MOLLUSCA : CLASS, CEPHALOPODA.

1. **Cephalopods.**—The only group of mollusks remaining, which we will study here, is that of the cephalopods—a group which contains the most highly organized animals among the mollusks. The name cephalopod is derived from two Greek words which mean feet on the head. To this class belong the octopus, cuttle-fish, squid, etc.

Fig. 107.—OCTOPUS.

2. **The Body covered with a thick Mantle.**—With the exception of one variety, cephalopods have no shell. The body is covered with a thick bag or mantle, which is

beautifully spotted, and which possesses the power of changing its color. The color is generally a mottled brown, but when irritated it changes to a reddish or purple hue, passing rapidly from one tint to another.

3. **The nearest Approach to a Brain.**—The head is distinct from the rest of the body, and contains nervous ganglia, protected by a covering of cartilage, which is the nearest approach we have seen to the brain of vertebrates. The large staring eyes are likewise more nearly perfect than any we have yet found.

4. **The Arms.**—The eight arms, or feet, whichever we choose to call them, surrounding the mouth are the most striking feature of the octopus (Fig. 107). They sometimes grow to a great length, and they have two rows of suckers on the underside (Fig. 108), which adhere so firmly to objects within their reach that these animals are dangerous foes. Fastening the suckers to their prey, they draw it down to their mouths, and hold it firmly until it is torn in pieces.

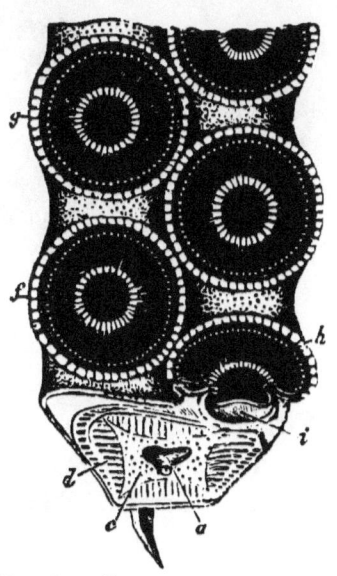

Fig. 108.—SUCKERS ON THE ARM OF A CUTTLE-FISH.

a, hollow axis of the arm, containing nerve and artery; *c*, cellular tissue; *d*, radiating fibres; *h*, raised margin of the disk around the aperture *f*, *g*, which contains a retractile membrane, or "piston," *i*.

5. **The Parrot's Beak.**—The mouth opens into a throat which is well supplied with implements for crushing food. In addition to a lingual ribbon, here are two large horny teeth, which from their shape are known as the "parrot's beak" (Fig. 109).

6. The Funnel.—The mantle is open at the neck, and expands to admit water to a chamber lying within, which contains the gills. The rim of the mantle then closes by powerful muscles, and the body contracts, and forces the water out in a jet through the "funnel." This funnel is a tube provided with a valve which closes after the water has escaped, and allows no water to enter through it from the outside.

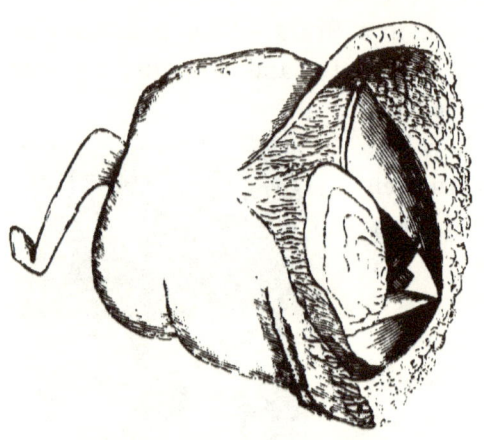

Fig. 109.—THE PARROT'S BEAK.

7. The Ink-bag.—Within the body is a sac containing an ink-like fluid which these animals throw out also from the funnel when they are alarmed. The surrounding water is thus discolored with a dense black cloud, and the octopods are enabled to escape from their enemies. This ink is sometimes used in water-color painting under the names of sepia and India-ink. The contents of the ink-bags obtained from fossil octopods have sometimes been dissolved, and still they yield sepia of a good quality.

8. Octopods found in most Seas.—Octopods are found in most seas, those living in mid-ocean sometimes being very large. Within the last few years some large specimens have been taken on the Newfoundland coast. Wonderful stories are told of octopods, but it is difficult to know how much of fable may have been interwoven with the truth. We at least know that they are active creatures, often jumping out of the water, and that they have a strange fashion of swimming backward.

9. **Manner of Propelling Themselves.**—Their only means of propelling themselves is by forcing water out of the funnel, the successive jets driving them backward, while the long arms trail uselessly after them. They also walk head-downward, with the rounded body above.

10. **Destroying Life.**—They spend much of the time partly concealed by the rocks, with their arms floating round in search of something to kill, for they are extremely greedy, destroying large numbers of fishes, crabs, and mollusks. Like the tiger, they seem to find pleasure in killing more than they need to eat. Their hiding-places are sometimes discovered by the number of dead shells scattered about. Octopods likewise are destroyed in large numbers by porpoises and whales.

11. **The Cuttle-fish.**—The cuttle-fish (Fig. 110) is much like the octopus, but it has two tentacles longer than the arms, with club-shaped ends. There are also narrow fins at the side of the body, and the

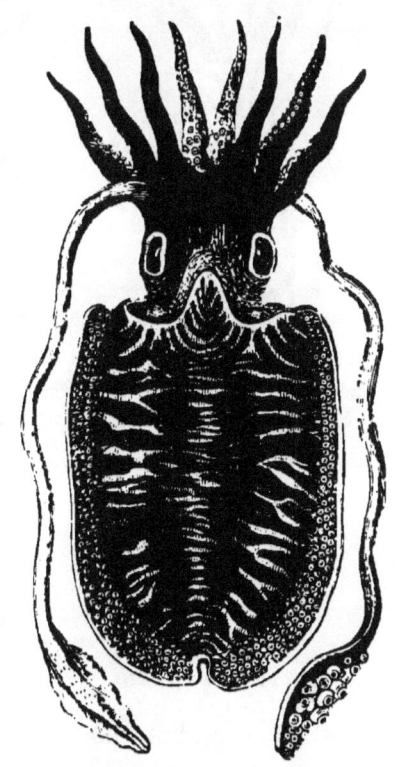

Fig. 110.—CUTTLE-FISH (ONE-FIFTH NATURAL SIZE).

mantle is supported on the inside by a thin plate which is known as the cuttle-fish bone. Cuttle-fish bones are commonly used in bird-cages. If you examine one carefully you will find it has no resemblance to true bone, being formed of layers, as shells are, with a hard covering. As the captive bird pecks at this it obtains small particles of lime, which substance is needed for forming its bones.

12. Cuttle-fishes do not lie concealed in caves waiting for their prey, but they come out boldly, and give their victims a fair chance.

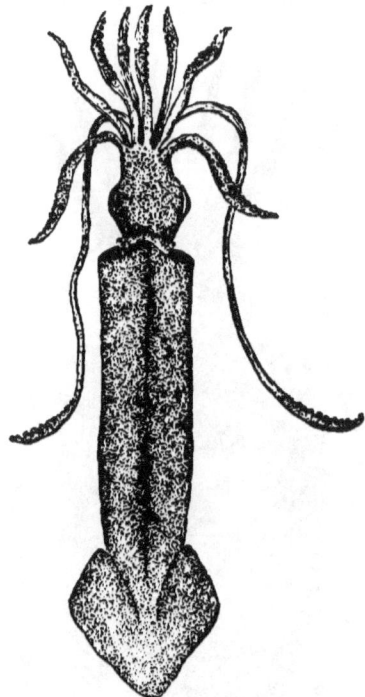

Fig. 111.—THE SQUID.

13. **The Eggs in Bunches.**— The octopus and cuttle-fish both attach their eggs by a cement secreted within their bodies to branches of seaweed, where they hang like bunches of grapes. The mother sometimes selects a snug retreat in the rocks for raising her young family, and barricading the entrance with pieces of rock or piles of shell, she allows no one to enter.

14. She is said to examine her eggs and rub them, sometimes syringing them with her funnel, as if to keep off parasites. In about five weeks the eggs are hatched. The little creatures are about the size of a grain of rice, with eight points from which the arms will grow, and they already have the power of changing their color. The mother is much exhausted after her long confinement, her nourishment the while not having been sufficient.

15. Used as Food.—In many countries these evil-looking creatures are used for food. They are highly esteemed on the shores of the Black Sea, for, as they are neither meat nor fish, they can be eaten during the long fasts of the Greek Church.

16. In Southern Italy the octopus is taken alive to market, and displayed in large tubs filled with sea-water. Here the animals writhe and twist their arms, and display their dread suckers. They all look straight in front of them with their great eyes, and at frequent intervals dis-

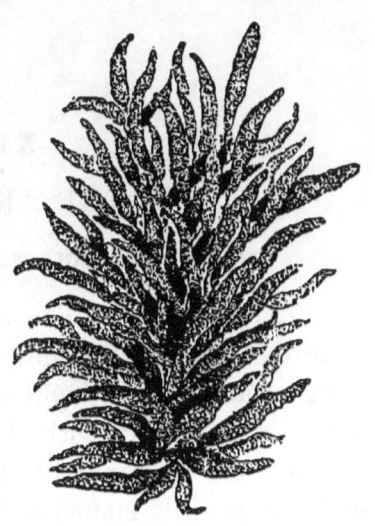

Fig. 112.—Egg-cluster of Squid.

charge water violently from their funnels in short, quick jerks. When a purchaser has selected one, the salesman seizes the octopus by the neck, and kills it by a skilful twist.

17. The Squid.—The squid (Fig. 112) is another one of the cephalopods. It is found in great numbers along the sea-shore, and is very generally used for bait in cod-fishing.

18. The eggs of the squid are enclosed within pod-shaped masses of stiff jelly, many of which are clustered together in one bunch. These pod-shaped pieces of jelly are sufficiently transparent to show the presence of many small eggs when held up to the light.

XXVIII.

THE ARGONAUT.

SUB-KINGDOM, MOLLUSCA : CLASS, CEPHALOPODA.

1. **The Argonaut.**—The argonaut, or paper-nautilus, is regarded as one of the most beautiful objects in the ocean. Who would have thought of finding a celebrated beauty in the same family with the disagreeable octopus! The charm must certainly be due to that lovely white shell which is prettily ribbed and fluted, and so transparent as to show the varying silver and rose tints of the body underneath.

2. **Resemblance to the Octopus.**—Notwithstanding these attractions, our eyes are now sufficiently trained to find in the argonaut many points of resemblance to the octopus. There are the unmistakable suckers on the arms, the great wide-awake eyes, and the curious funnel projecting beyond the shell just below them.

3. **The Shell secreted by the broad Arm.**—In addition to these, there are many new points of interest. You will notice in the upper figure of this picture that the two hinder arms are spread out into flat, sail-like membranes, which here only partly cover the shell. They may, however, be extended so as to cover it entirely. Indeed, the shell has been secreted by these broad membranes, and if it is broken in any way, the injuries are soon repaired by new shelly matter deposited just where it is needed.

Fig. 113.—ARGONAUTS.

4. The Body not fastened to the Shell.—Although the argonaut lives in this shell, its body is nowhere fastened to it, neither does it fit the shell and fill it up, as other mollusks do. It merely sits in the graceful shell as in a boat, and holds on by its webbed arms.

5. Fanciful Stories.—Fanciful stories have been told of the argonauts, and persons were led to believe that they sailed over the waves, with their webbed arms held aloft as sails to catch the breezes, and that their straight arms were used as oars. So far from sailing in this fantastic fashion, the argonaut rarely comes to the surface, but passes its days in deep water or upon the bottom of the ocean. Here it crawls head-downward, with its shell over its back, using its arms in place of feet.

6. In reality the argonaut swims just as its relatives do —by squirting itself backward. Gathering the arms together in a straight line, as shown near the middle of Fig. 113, it takes in sea-water under the mantle, and forcibly expels it from the funnel.

7. How snugly the lowest one in the picture has tucked itself away in the shell! Still, it has an eye for all that goes on around it.

8. A Dainty Shell for a Cradle.—The charming part of our story has yet to be told, for you must know that these dainty shells are merely nests, with which the females are provided to protect themselves and the bunches of eggs which they carry. The young ones are hatched in this lovely floating cradle, and are thus shielded from many dangers to which they would be exposed in the open sea. In the middle figure you may see the large bunch of eggs on top of the shell and partly concealed by the body of the parent.

9. The Male Argonaut.—The male argonaut is very unlike the female. Not being more than an inch in length, and having no shell, it was not recognized until quite recently as the mate of the handsome paper-nautilus.

10. **Habits Unknown.** — These animals live in tropical seas, but their shells have sometimes been washed on our own shores. It is impossible to know the habits of such deep-sea dwellers, since their haunts are completely hidden from our view.

11. Argonauts have interested thoughtful men from a very ancient date. Their appearance on the water was welcomed as an indication of fine weather, and one of the Greek poets long ago wrote, "O fish, justly dear to navigators! thy presence announces winds soft and friendly; thou bringest the calm, and thou art the sign of it."

XXIX.

THE PEARLY NAUTILUS.

SUB-KINGDOM, MOLLUSCA: CLASS, CEPHALOPODA.

1. **The Pearly Nautilus.**—The most interesting of all the cephalopods is perhaps the pearly nautilus. Unlike other members of its class, this animal is supplied with a true external shell, which is divided into many chambers; hence, it is often called the "chambered nautilus."

2. **The Chambered Shell.**—In its natural condition the outside of the shell resembles white porcelain streaked with reddish-brown stripes. The nautilus shells usually seen in cabinet collections have been polished; this outside striped coating has thus been removed, and nothing remains but the lustrous pearl underneath.

3. The shell is elegantly shaped and proportioned, but gives no hint of the curious arrangement inside until it is cut open. It is then found to contain many chambers partitioned off by curved, pearly plates which you can readily see in Fig. 114. The animal always occupies the outer and larger chamber, as here represented, retiring from it in its turn and walling it up as the shell increases in size to meet the needs of the growing body.

4. In this way each chamber has been successively the home of the nautilus, and has been abandoned when it ceased to be desirable.

" Year after year beheld the silent toil
  That spread his lustrous coil ;
    Still, as the spiral grew,
    He left the past year's dwelling for the new,
    Stole with soft step its shining archway through,
  Built up its idle door,
  Stretched in his last-found home, and knew the old no more."

5. **The Siphuncle.** — There is a curious tube, or "siphuncle," extending from the body through all the chambers to the end of the coil. Its use is not positively known, although it may be instrumental in compressing the gas

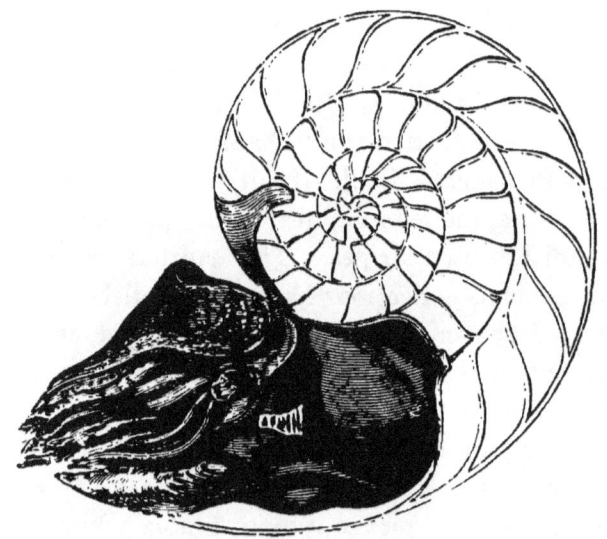

Fig. 114.—PEARLY NAUTILUS, WITH SHELL CUT OPEN
(ONE-HALF NATURAL SIZE).

with which the chambers are probably filled, thus affecting the weight of the shell, and enabling the animal to rise or sink in the water when it wishes.

6. **Means of Swimming.**—Our beautiful nautilus has discovered no more graceful means of swimming than by ex-

pelling water from the funnel, as others of its family do, but it has not their peculiarity of squirting ink, insomuch as it possesses no ink-bag.

7. **The Organs.**—It has many short arms, which are highly sensitive, but which have none of the suckers so remarkable in the cuttle - fish. The mantle is thickened into a leathery fold or hood over the head, which closes the shell when the animal retires within it.

8. The mouth is surrounded by a fleshy lip and several additional folds, and it opens into a cavity where the parrot's beak and the lingual ribbon are situated. The eyes are attached by short stalks to the sides of the head.

9. **Living Specimens rare.**—Although nautilus shells are quite common, only a few specimens of the animal have ever been obtained ; from which fact it is inferred that the nautilus lives only at great depths, in tropical regions of the Pacific and Indian oceans. Unfortunately, we know almost nothing of its habits.

10. **The Last of its Race.**—The nautilus is especially interesting, since it is the last member of a once numerous race of four-gilled cephalopods with external shells, which formerly occupied the seas. Entire families have ceased to exist, and are known to us only by fossil remains, which are very abundant in the rocks, more than two thousand species being known.

Fig. 115.—AMMONITE.

11. **Ammonites.**—Among the most interesting of these are the ammonites, one of which is shown in Fig. 115. Their chambered shells are much like nautilus shells, but instead of having partitions with plain edges, the partitions are folded and

crinkled, forming curious patterns on the outside of the shell.   Ammonites evidently lived in the deep sea.   They are found of all sizes, varying from an inch to more than a yard in diameter.

12. These ancient four-gilled forms have been succeeded by the two-gilled cephalopods (such as the octopus and cuttle-fish) without shells, which now monopolize the ocean. The beautiful nautilus has gradually decreased in numbers, and will probably become extinct also, as the rest of its family have done.

XXX.

BACKBONED ANIMALS.

SUB-KINGDOM, VERTEBRATA.

1. The Backboned Family.—An important point in our studies is now reached, and we are about to enter upon that great sub-kingdom which is spoken of as the "backboned family." The animals comprised in this large family differ so greatly from the Radiates, Articulates, and Mollusks, which we have been studying, that it seems necessary to pause here and examine some of their peculiarities.

2. We shall find among the backboned animals a great variety of forms and habits ; still, we can trace in their physical structure an unbroken series, and passing regularly up from the lower forms of this type, we shall gradually approach animals that are highly endowed with intelligence and strength. Fishes, we know, live only in the water. So also do frogs and toads during the early part of their lives, but after

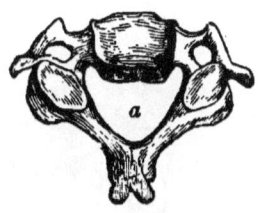

Fig. 116.—A VERTEBRA.
a, spinal canal.

they become fully matured they mostly live on the land, and have many points of resemblance to reptiles. In turn, the reptiles lead on to birds, which, by their peculiar strength and lightness, are fitted to live almost wholly

in the air.   Last of all are the mammals, whose superior
endowments crown the list.

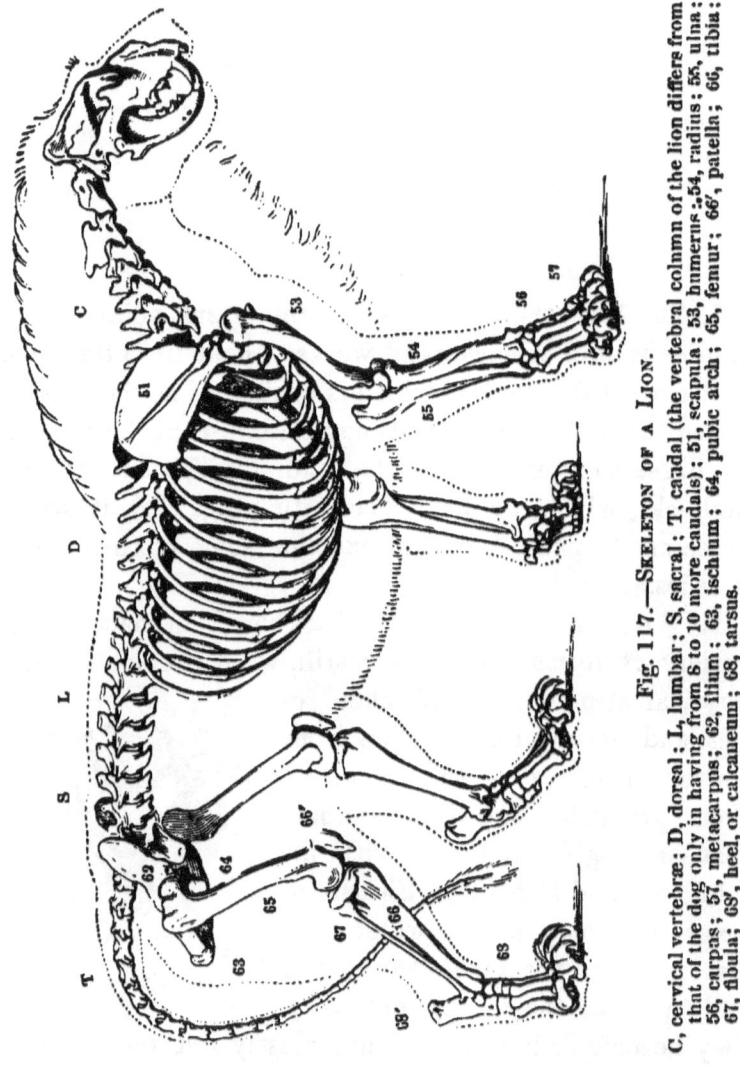

Fig. 117.—Skeleton of a Lion.

C, cervical vertebræ; D, dorsal; L, lumbar; S, sacral; T, caudal (the vertebral column of the lion differs from that of the dog only in having from 6 to 10 more caudals); 51, scapula; 53, humerus; 54, radius; 55, ulna; 56, carpus; 57, metacarpus; 62, ilium; 63, ischium; 64, pubic arch; 65, femur; 66′, patella; 66, tibia; 67, fibula; 68′, heel, or calcaneum; 68, tarsus.

3. **The Spinal Column.**—The most important peculiarity
of this family is that the greater part of its members
possess a " backbone," or, more properly speaking, a spinal

column, which is composed of a chain of small bones or
vertebræ. Owing to this fact, the name *Vertebrata* has
been given to the sub-kingdom. The vertebræ are united
side by side by means of ligaments, and as each vertebra
has an opening through its centre, as
is shown at *a*, in Fig. 116, the chain
forms a continuous canal throughout
the entire length of the spinal column.

4. **The two Tubes of Vertebrates.**—In
the spinal canal, which we have just described, lies the spinal cord, safely encased in bone, and connecting with the
brain through an opening in the lower
part of the skull. You will observe,
therefore, that the brain and the spinal
cord, the large masses of the nervous
system, are shut off by a special tube
from other parts of the body.

5. Carrying this idea yet further,
we will now consider the main cavity
of the body, which (as in the lion, Fig.
117) is formed by the ribs, the backbone, and the breastbone, as a second
tube for containing the heart, lungs,
stomach, etc.

6. **The distinction between Vertebrates
and Invertebrates.**—This especial provision for the nervous system is the
great distinction between vertebrates
and those animals having no spinal col-

Fig. 118.— HUMAN
BRAIN AND SPI-
NAL CORD SEND-
ING OFF NERVES.

umn, which are called invertebrates. The body of invertebrates may be looked upon as one single tube, in which
the nervous system is not separated from other organs;

whereas, the body of vertebrates consists of two distinct tubes, one for the large, nervous masses, the other for the organs of digestion, circulation, etc.

7. **The Nervous System.**—The brain, lying within the skull, and the spinal cord proceeding from it, are the great centres which give rise to the symmetrical pairs of nerves passing to all parts of the body (Fig. 118). These nerve-

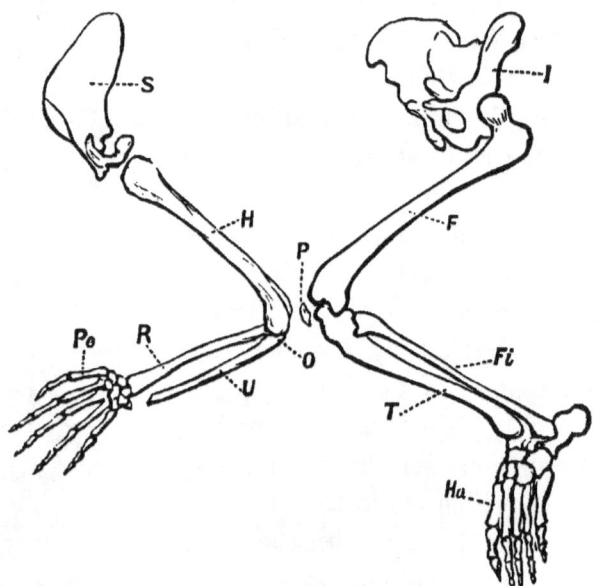

Fig. 119.—ARM AND LEG OF A MAN AS THEY ARE SEEN WHEN ON ALL-FOURS.

S, scapula; I, ilium, or shinbone of pelvis; H, humerus; F, femur; O, olecranon, or tip of the elbow; P, patella; U, ulna; T, tibia; R, radius; Fi, fibula; Po, pollex, or thumb; Ha, hallex, or great toe.

centres are, as we have seen, well protected by the wonderful chain of bones constituting the spinal column. In addition to this arrangement, there is a series of nerves supplied to the thorax and abdomen known as the sympathetic nerves, which regulate the digestion, respiration, and the circulation of the blood.

8. **The Skull.**—That strong, bony box which we call the

skull not only contains the brain, but it also protects the delicate organs of sight and of hearing, as well as those of taste and smell, all of which are lodged in its bony cavities.

9. **The Skeleton.**—Another peculiarity of vertebrates is that they possess a jointed skeleton which is always internal. The hard bones composing this skeleton are not dead and lifeless, as they look to be, but they undergo a continual change, since they are nourished by the blood which, in its circulation through the bony tissues, carries off all the waste particles, and deposits new materials for their growth and repair.

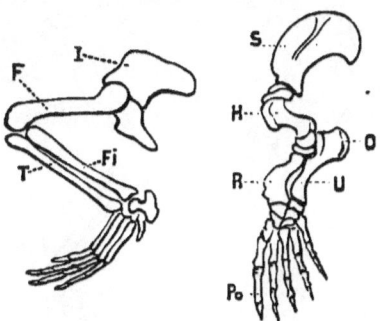

10. **The Limbs.**—Animals of this family never have more than two pairs of limbs, and these are jointed and turned away from the back. By far the greater number of vertebrates are supplied with both pairs

Fig. 120.—Hind-leg of Alligator and Fore-leg of Seal.

S, scapula; I, ilium, or shinbone of pelvis; H, humerus; F, femur; O, olecranon, or tip of the elbow; U, ulna; T, tibia; R, radius; Fi, fibula; Po, pollex, or thumb.

of limbs; still there are some animals, as for instance the whale, in which the limbs are only partially developed, and others again, like the snake, in which the limbs are altogether wanting.

11. As the limbs of different animals are employed for a great variety of purposes, we shall find that they are curiously modified to serve these various uses, yet a similarity may be observed in the structure of all. Thus the fore-legs of quadrupeds are represented by the arms of man, by the wings of a bird, and by the swimming-pad-

dles of a whale. If you should compare the arm of a man with the leg of a seal or the wing of a bat, as represented in the accompanying figures, you will discover a close resemblance between them.

12. **The Digestion of the Food.**—All vertebrates have a mouth, which generally is furnished with teeth. The food is mostly cut and divided in the mouth and mixed with saliva, after which it is swallowed and digested, and the nutritious portions are absorbed into the blood.

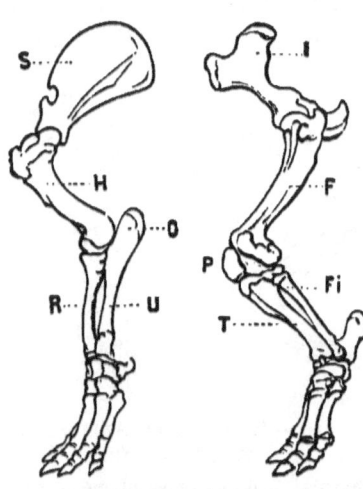

Fig. 121.—FORE AND HIND LEG OF A TAPIR.

S, scapula; I, ilium, or shinbone of pelvis; II, humerus; F, femur; O, olecranon, or tip of the elbow; P, patella; U, ulna; T, tibia; R, radius; Fi, fibula.

13. **The Heart.**—The heart of all vertebrates contains at least two chambers, and in the higher classes of animals it has four distinct chambers. These chambers are supplied with valves at their openings which allow the blood to pass through, but which close in such a manner as to prevent its return in the opposite direction.

14. **Circulation of the Blood.** —Dr. Harvey, in 1619, was the first person that taught the great fact of the circulation of the blood, and it is now so well understood as to attract but little attention. We know that the blood of living animals is continually flowing to every part of the body through closed tubes, or blood-vessels, as they are called, the arteries being employed in carrying it from the heart, and the veins in returning it again.

15. **The Blood.** — We are accustomed to think of the

blood simply as a red fluid, whereas, upon examination, it is found to be a clear liquid, almost without color, in which floats a multitude of minute particles or "corpuscles," so exceedingly small that they can be discovered only with a powerful microscope. Some of these corpuscles are red, others are white. The red corpuscles have a tendency to run together into piles like buttons on a string, and they are so numerous as to tinge the blood with their red color (Fig. 122).

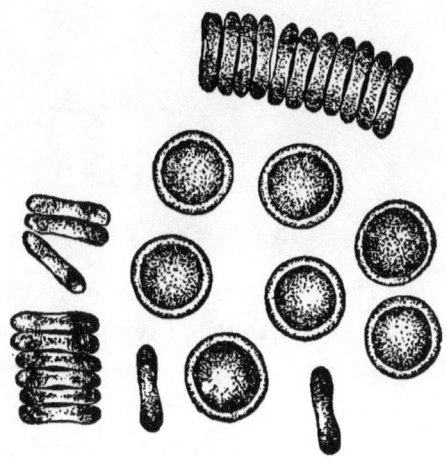

Fig. 122.—Blood Corpuscles of Man.

16. **The Breathing of Vertebrates.**—We shall find a great difference among these animals with regard to their manner of receiving a supply of fresh air. Fishes and amphibious animals are furnished with gills, and derive their supply of air from the water, but the higher vertebrates breathe by true lungs, and the process of airing the blood is greatly assisted by the action of the skin.

XXXI.

FISHES.

SUB-KINGDOM, VERTEBRATA : CLASS, PISCES.

1. **Fishes.**—It is not an easy thing to study the every-day life of fishes. Living as they do in the water, and keeping out of sight, our only hope of observing them is from an occasional glimpse, which gives us little oppor-tunity to learn their habits and peculiarities. Preserved specimens are of no great help; they serve, however, to remind us that much of the charm of fishes lies in the grace of their movements and in the delicate lustre which plays upon their sides as they glide through the water, but which is lost soon after death.

2. **Fishes are well adapted to Swimming.**—The shape of fishes is such as to admit of their swimming easily and smoothly through the water with.the least possible fric-tion. They are further aided in swimming by their smooth, slimy coating, which generally consists of scales overlapping one another like tiles on a roof.

3. Dr. Hartwig says of fishes: "We wisely endeavor to imitate this peculiar form in the construction of our ships, yet the rapidity with which the fastest clipper cleaves the waters is nothing to the velocity of an animal formed to reside in that element. The flight of an arrow is not more rapid than the darting of a tunny, a salmon, or a gilt-head through the water. Every part of the body

Fig. 123.—NEST OF THE SUN-FISH.

seems exerted in this despatch: the fins, the tail, and the motion of the whole backbone assist progression; and it is to this admirable flexibility of body, which mocks the efforts of art, that fishes owe the astonishing rapidity of their movements."

4. **The Vertebræ.**—On examining the backbone of a fish, you will find it to consist of circular vertebræ, which are concave at each end. The space between the vertebræ is filled with a jelly-like substance, giving great freedom of motion. The ribs are not much curved, and are attached to the spinal column at one end.

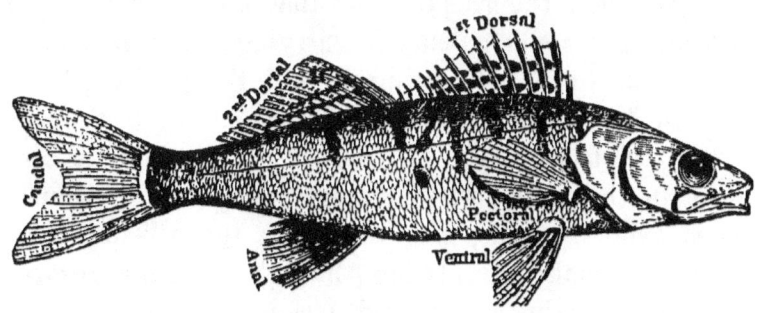

Fig. 124.—THE FINS OF A FISH (PIKE-PERCH).

5. **The Fins.**—Notice the two kinds of fins which are represented in Fig. 124. The "paired fins," which are arranged on opposite sides of the body, are the true limbs of the fish. Those near the gills, named the pectoral fins, represent the fore limbs of quadrupeds, while the ventral fins, which are lower down, and which may sometimes be wanting, represent the hind limbs. Besides these paired fins there are other single fins upon the middle of the back and underneath the body.

6. The fins are chiefly used to balance the fish in an upright position, while the principal swimming organ is the

tail, which is set vertically at the end of the spine, so as to work from side to side.

7. **The Swim-bladder.**—As the weight of the body is greater than that of the surrounding water, most fishes are supplied with a "swim-bladder" (Fig. 125, *vn*) which connects with the œsophagus, and which, being filled with air, assists the fish to rise or sink in the water.

8. **The Gills.**—Fishes' gills are leaf-like bodies lying in cavities on each side of the neck (Fig. 125, *br*), and covered by plates called gill-covers; they may often be seen gently moving when a fish is in its native element. You may also notice the gills in fishes that have been arranged on a string for convenience in carrying; in such cases it is customary to put the string through the "gill-slit."

9. **The Manner of Breathing.**—The breathing of fishes is a very simple process. All the air they require is contained in the water, which enters freely at the mouth, passes over the gills, and escapes at the gill-slit. The blood continually circulating through the gills absorbs oxygen from the water and becomes purified. This breathing from water resembles the act of swallowing, with the important difference that the water passes to the gills, and not to the stomach.

10. We sometimes say, when a fish is taken out of the river, that it dies for want of water. Strictly speaking, it would be more correct to say that it dies for want of air. Surrounded as it is by air, the fish can make no use of it, because it is not mixed with water, and the poor creature flounces and throws itself uneasily about as it slowly suffocates.

11. **Circulation of the Blood.**—The heart of fishes has but two cavities, an auricle and a ventricle, which are shown in the ideal plan of the circulation of fishes (Fig. 126), at

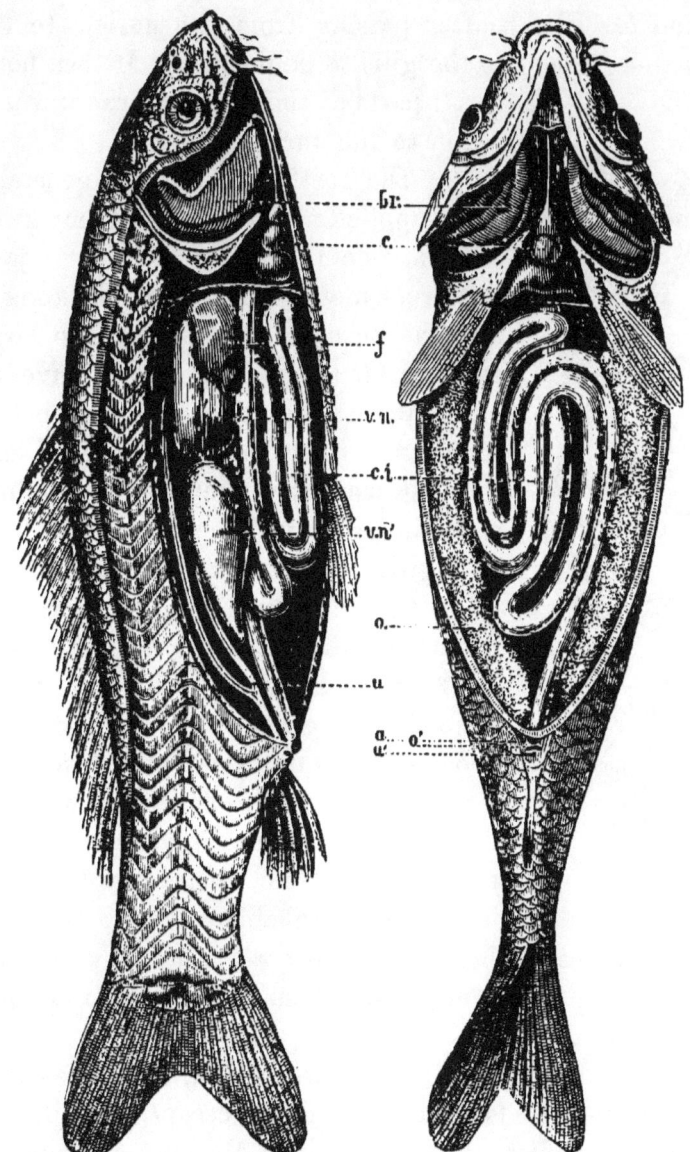

Fig. 125.—ANATOMY OF THE CARP.

br, branchiæ, or gill-openings; *c*, heart; *f*, liver; *vn, vn'*, swimming-bladder;
ci, intestinal canal; *o*, ovarium; *u*, ureter; *a*, anus; *o'*, genital opening;
u', opening of ureter. The side-view shows the disposition of the muscles in
vertical flakes.

a and *b*. Blood, after passing from the auricle to the ventricle, is sent to the gills to be purified. It then flows to all parts of the body before returning again to the auricle.

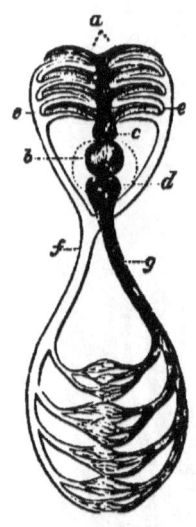

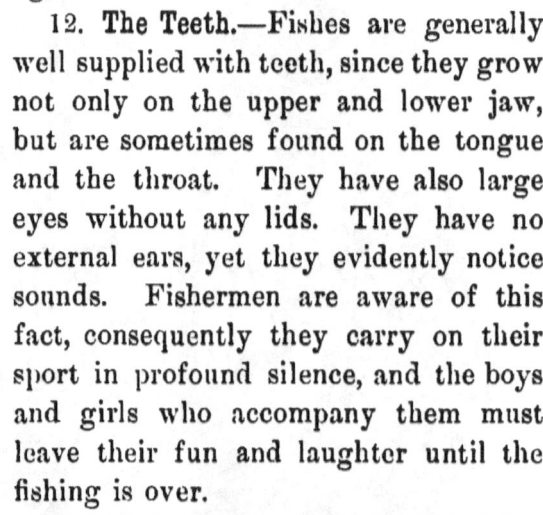

Fig. 126.—PLAN OF CIRCULATION IN FISHES.

a, auricle; *b*, ventricle; *c*, branchial artery; *e*, branchial veins, bringing blood from the gills, *d*, and uniting in the aorta, *f*; *g*, vena cava.

12. **The Teeth.**—Fishes are generally well supplied with teeth, since they grow not only on the upper and lower jaw, but are sometimes found on the tongue and the throat. They have also large eyes without any lids. They have no external ears, yet they evidently notice sounds. Fishermen are aware of this fact, consequently they carry on their sport in profound silence, and the boys and girls who accompany them must leave their fun and laughter until the fishing is over.

13. **Their Enemies.**—Although fishes seem to have a happy time as they dart about in the water, the truth of the matter is they live in continual warfare, first pursuing their prey, and then, in turn, flying from their own enemies. They have probably more to suffer in this way than other animals, for a great many enemies are waiting to pounce upon them, their eggs, and their young ones.

14. **Flying-fishes.**—Flying-fishes (Fig. 127) often leap into the air in large companies to escape pursuit, and their blue bodies and silvery wings glisten prettily in the sunlight. Even here they sometimes meet with new dangers from the greedy gulls and other sea-birds, so they find safety neither in the water nor in the air. The large

fins of these fishes act like wings, and enable them to take long, low leaps into the air, but they have no power of raising themselves after having once left the water.

Fig. 127.—FLYING-FISH.

15. **The Eggs of Fishes.**—The eggs of fishes are tiny affairs, covered with a thin skin, so transparent that the young fish may be seen tumbling around inside for a day or two before it is hatched. When the baby-fishes first leave the egg they swim about for some time with the yolk-bag hanging underneath the body; they take no food during this time, but are nourished by the oily contents of the yolk-bag.

16. Fishes produce large quantities of eggs, a single cod-roe, for instance, having been found to contain nine million eggs; but, as we have just seen, a very large proportion of the young fishes are devoured; being helpless little creatures, they fall an easy prey to their larger neighbors.

17. **Care of the Young.**—Generally, fishes take no care of their eggs or their young ones. There are exceptions,

however, to this rule, and some kinds of fishes prepare nests in the bottom of streams, like the sunfish at the beginning of this chapter, while the stickleback builds a true nest of grass and weeds fastened together with the sticky slime of his own body (Fig. 128). There is a hole entirely

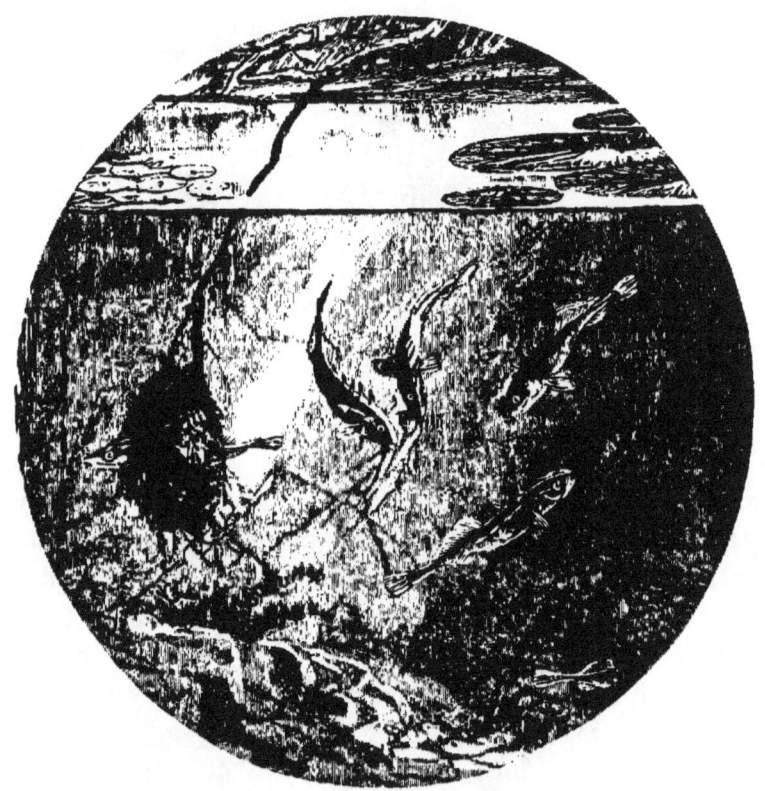

Fig. 128.—NEST OF THE STICKLEBACK.

through the nest, from one side to the other, that water may constantly flow over the eggs. The male defends the nest bravely, for, strangely enough, among fishes it falls to the lot of the fathers to build the nest and care for the young ones.

18. These families are mostly large, and you may imagine it is no easy matter to keep the active little ones together when swimming is so natural to them. The attempt to protect them leads to some curious habits; as, for instance, the habit which many fishes possess of taking their young ones into their mouths for safety. In times of danger the father opens his wide mouth, and the whole swarm rushes into the temporary asylum thus provided.

19. **The Sea-horse.**—The sea-horse (Fig. 129) has a novel way of protecting his precious infants, for he gathers them all into a curi-
ous pouch on the lower part of his body which is especially provided for this purpose. The young fishes are taken into the pouch as soon as they are hatched, and snugly carried there

Fig. 129.—THE SEA-HORSE.

until they are old enough to take care of themselves. What an odd sight it must be when these spry little creatures are first turned out of their cradle into the wide ocean, and the whole swarm starts off to see the world.

20. One would scarcely suspect this odd-looking sea-horse, with its long snout, of being a fish. It has a singular habit of twisting its tail round some branch of sea-weed and standing upright in the water, as if watching all that takes place around it. Being a poor swimmer, it often floats with the sea-weed for long distances in this erect position.

21. **Brilliant Coloring of Tropical Animals.**—Tropical birds and flowers, we know, are brilliant in color; so also

9*

are the inhabitants of tropical oceans. We noticed this peculiarity in jelly-fishes and in shells, and the same is true with regard to fishes. Some of the gayest fishes live among the coral reefs. The warm waters in which the coral polyps thrive and spread their flower-like tentacles to the sun are further enlivened by glittering fishes, which glide in and out among the brilliant coral branches, and remind us of the similar fact that dazzling birds hover over the brightest flowers.

Fig. 130.—SHARK.

22. **Selecting Mates.** — We might suppose that these lowly creatures would not pay much regard to beauty in selecting their mates, but with fishes as with every other species of animal there are points of difference, which we would probably not notice, but which lead to the selection of certain individuals in preference to others. It has been observed that many fishes grow brilliant as the season approaches for mating. All members of the trout family, for example, are arrayed in their brightest colors during spawning-time.

23. **Sharks.**—Very unlike the graceful fishes we have been examining are those tyrants of the ocean, the sharks, which are by far the largest and strongest of the fishes. The stout body is well shown in the picture (Fig. 130), and you must not fail to notice how odd the tail is. The top point is much longer than the lower one, whereas the tails of most fishes nowadays are even. Sharks have several other peculiarities which cause them to be classed with an ancient race of fishes, few of which are now living. Their skeletons do not consist of bone, but of hard gristle. Instead of a scaly covering, their skin is set with hard knobs, and the slits on the side of the neck take the place of gills.

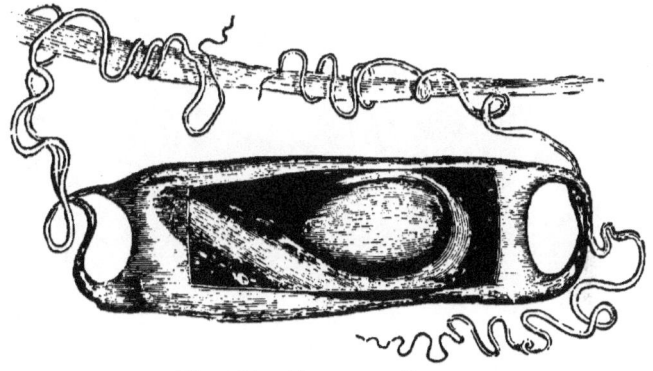

Fig. 131.—EGG OF A SHARK.

24. **The Mouth.**—The mouth, you see, is on the underside of the head, and in consequence of this arrangement sharks are usually obliged to turn over before biting. A savage-looking mouth it is, too, with several rows of sharp teeth pointing backward. These teeth are never fixed in sockets, however, but are merely imbedded in the lining of the mouth.

25. **Sharks' Eggs.**—The female shark lays but two eggs, which are enclosed in leathery, purse-shaped cases. The

four corners of the case are lengthened into tendrils,
which, becoming entangled in sea-weed, hold the egg in
place.   Part of the case has been removed from the egg
which is shown in Fig. 131, that we may see, within, the
young fish with the yolk-bag attached to it.   The empty
black cases of sharks' eggs are often picked up on the sea-
shore, and the sand which rattles out of the hollow case
may assure you that its former occupant has already
escaped.

XXXII.

THE MIGRATION OF FISHES.

1. **Migration of Fishes.** — The curious habit possessed. by some animals of moving in companies from one place to another at certain seasons of the year is spoken of as "migration." We are probably most familiar with the migrations of birds, but many kinds of fishes yield to the same instinct, and their migrations are closely connected with the production of their eggs.

2. A good illustration is furnished by cod, mackerel, and herring, all of which select shallow water near the coast for depositing their eggs, and approach the shore for this purpose in enormous shoals, or schools, as they are called. In these migrations the fishes are crowded so close together as almost to force one another out of the water, and they are pursued by many birds and marine animals, in their efforts to escape from which they are often washed ashore in masses.

3. **Busy Times among the Fishermen.** — The arrival of these schools upon the coasts causes busy, bustling times among the fishermen, whose boats may then be seen hovering over them like great flocks of sea-birds, anxious to catch all they can while the harvest lasts. Mackerel-fishing is thought to be fine sport, and is performed under full sail. The faster the boat moves the better the mackerel bite. They rush after the bait as if mistaking it for

escaping prey, and as the boat glides through the great shoals of fish, all hands on board are kept busy hauling in the lines and putting on fresh bait.

Fig. 132.—A Fishing Fleet.

4. The Migration of Salmon.—The migrations of salmon are especially interesting. These fishes, although hatched in fresh water, pass the greater part of their lives in the ocean, and at certain seasons they ascend the rivers in large companies to deposit their eggs. It is believed that they return year after year to the same locality; so the baby salmon are raised in the old home of their parents, who, nevertheless, have become in the mean time great rovers.

5. **Efforts to reach the Source of the River.**—Young salmon cannot live in salt water, consequently the eggs must be placed where there is little danger of the young fishes drifting out to sea, and upon these journeys the impulse of the parents is so strong to reach the source of the river that they seem determined to overcome all obstacles.

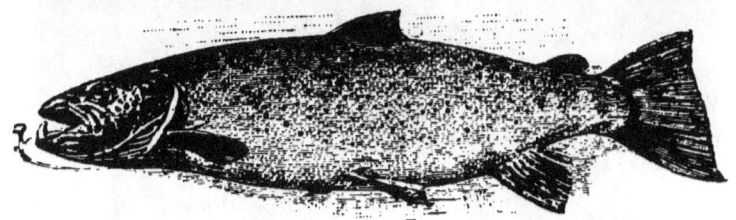

Fig. 133.—MALE SALMON.

They even leap the water-falls, and in doing this they display great perseverance. In leaping they throw the body into a curve, resting on the head and tail ; they then make a sudden spring, which is greatly aided by the pressure of the tail upon the water. The first attempt is often unsuccessful, and they fall, perhaps, upon the rocks or upon the bank of the river ; not discouraged by the failure, they struggle back to the water and try again.

6. It is now customary to place fish-stairs where there is a water-fall or a dam too high for the salmon to mount. These consist of a series of steps made of wood or stone, which divide the height into a succession of small falls. The salmon soon learn to leap from one step to another, and in this way they reach the top without difficulty.

7. **Spawning.**—Salmon, it is said, have a leader on these journeys, and follow him in regular order. Having ar-

Fig. 134.—SALMON-FISHING.

rived at some suitable place, they hollow out nests in the bottom of the stream, and deposit their eggs, covering them with gravel, and then taking no further care of them.

8. They eat little or nothing while in fresh water, and they reach the spawning-ground bruised and exhausted by the hardships they have endured. They therefore rest for a while after the spawning process, which occupies eight or ten days, and then eagerly return to the sea.

Fig. 135.—FEMALE SALMON AFTER SPAWNING.

9. **The Young Salmon.**—The eggs left under the gravel finally hatch out, and the young fishes work their way slowly down the river, to make their first visit to the sea. These fishes increase but little in size while in fresh water, whereas in the ocean the rapidity of their growth is almost incredible.

10. **The Shape and Color.**—Salmon are remarkably graceful fishes, and their tapering shape is well suited to rapid motion. Their bluish-gray color shades into a silvery-white underneath, and the upper part of the body is marked with black spots. As breeding season approaches, they not only grow brilliant in color, but a change still more remarkable than this takes place in the mouth of the males. The under-jaw forms itself into a strong hook, which may be seen in Fig. 133. This hook is used in the fierce combats between the males at that season, and it often inflicts deadly wounds.

11. **The Delicacy of the Flesh.**—The pink-tinted flesh of

the salmon is exceedingly delicate, and probably owes its peculiar flavor to the eggs of echinoderms and crustaceans, of which this fish is especially fond.

12. Salmon-fishing.—Some of the British and Norwegian rivers contain celebrated salmon-leaps, and they are visited by many persons who enjoy the sport of salmon-fishing. The proper time for catching these fishes is when they ascend the rivers to spawn, for they are lean and poor on their return. At this time, however, they are ravenously hungry, and prove a serious annoyance to anglers, whose artificial flies are designed to attract only the good fishes fresh from the ocean. Salmon are also caught with nets and weirs, and with the spear.

13. The Cultivation of Fishes.—Much attention has recently been paid to the cultivation of fishes, which is merely the revival of an old art. A " fish - farm " consists of a set of troughs, standing each one a little higher than the next in the series, with fresh water constantly flowing through them. In these troughs fish-eggs are hatched by artificial methods, and when the young fishes have grown to a suitable size they are successfully planted in our rivers and streams.

XXXIII.

FROGS AND TOADS.

SUB-KINGDOM, VERTEBRATA : CLASS, AMPHIBIA.

1. **Frogs.**—Most of you, perhaps, already know that the funny little tadpoles in our ponds and ditches turn into frogs. Let us notice the remarkable changes which take place before tadpoles can pass in this way from the life of a fish to that of a land animal.

2. **Frog Spawn.**—We will begin with the eggs, which are little black specks not larger than shot, scattered through a lump of clear white jelly.

Fig. 136.—THE FROG.

This mass is called "frog spawn," and it is mostly attached to sticks or grass in the water near shore (Fig. 137). The jelly holds the eggs together, that they may not drift away, and it also supplies nourishment to the young animals when first hatched.

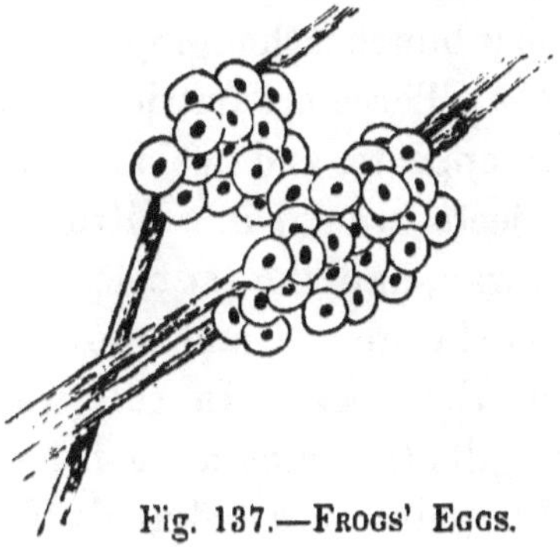

Fig. 137.—FROGS' EGGS.

3. **Tadpoles.**—If you should gather some of this frog spawn in the spring, and put it in a vessel of water with a few water-plants, you will have good entertainment for several weeks. First the round black specks begin to lengthen, then soon to wriggle about. Gradually the jelly mass disappears, and the young tadpoles, with big black heads, dart hither and thither, rapidly wagging their long flat tails as they swim through the water—a sight with which all country children are familiar.

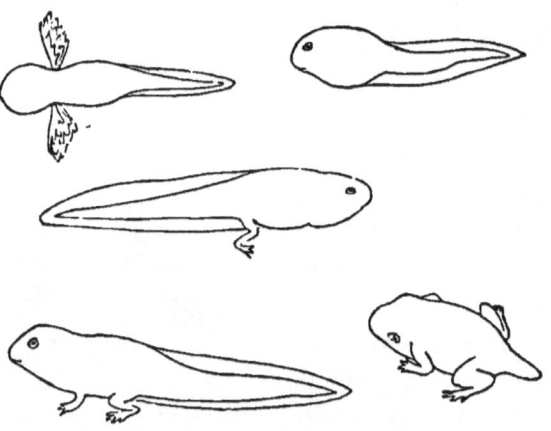

Fig. 138.—From a Tadpole to a Frog.

4. **Changes Tadpoles undergo.**—When they grow a little larger you can discover feathery bunches hanging at the sides of the head, as in Fig. 138, *a*. These are outside gills. After a time the wide mouth appears, and we find the tadpole trying to nibble at objects around it. Little by little the outside gills shrink away, and the tadpole then breathes by taking water in at the mouth and allowing it to run out through slits in the neck. In this way the water passes over internal gills the same as in fishes. Indeed, there is but little at this point in a tadpole's his-

tory to distinguish it from a fish, and it bears very slight resemblance to the form it is soon to develop.

5. Eyes and nostrils now make their appearance (Fig. 138, *b*), and soon two little lumps come on the sides, which will grow some day into hind-legs (*c*). The front legs do not show until later, and then the tadpole is well supplied with limbs, having four legs and a broad swimming tail, as you see at *d*.

6. The odd creature will now be found spending much time at the surface, with its mouth out of water, for it is trying still another plan for breathing.

7. While these changes have been taking place on the outside of the animal, still more important changes have been going on within its body. Lungs have been growing, and as the tadpole accustoms itself to breathing with the new lungs, the blood gradually changes its course, and rushes to them to be purified, instead of going to the gills as before. Consequently, the internal gills are no longer needed, and they also shrink away.

8. The young tadpole had at first a true fish's heart, with only two chambers, but now a third chamber grows, and we have our first instance of a three-celled heart.

9. **The True Frog.**—This active little creature now deserves the name of frog (Fig. 138, *e*). It swims with its new legs, and takes such long leaps that you must keep a close watch or it will jump out of your artificial pond and escape further observation. As the tail is no longer needed, it shrivels away little by little, like the gills, until there is no trace of it left.

10. When they have reached this period, frogs, in their native home, are ready to hop boldly on shore, although most of their time is passed in the water, perched on some stick or stone. When cold weather comes, they drop to

the bottom of the pond, and spend the winter in a torpid state.

11. Skeleton of a Frog.—Do you see in the frog's skeleton (Fig. 139) how much longer the hind‑legs are than

the front ones? This arrangement answers very well for leaping, and those long toes are usually joined with a web to assist in swimming.

12. The Breathing.—You will also notice that there are no ribs, so the frog cannot breathe

Fig. 139.—Skeleton of Frog.

as we do. Our ribs are raised each time we breathe, and the air rushes in through the nose and mouth to fill the empty space thus made in our chests. But as the frog has no ribs by which to enlarge its chest, it simply closes its lips and swallows the air which is in its mouth. A frog has no other way of breathing, and it is possible to suffocate one by fastening open its mouth.

13. The Tongue.—The long tongue of these animals is fastened at the front of the mouth, and the sticky point is turned over so that it can dart forward instantly, then fold back to snap up living insects.

14. The Development of Toads.—The history of toads is like that of frogs, except that their eggs are laid in long strings of jelly (Fig. 140), which may be found floating on ponds and ditches in the spring. As their young ones can live only in water, these animals lay their eggs either in the water or on trees and plants overhanging a pond, into which they are washed by the rain. Large numbers of toads thus come to perfection about the same time, and are ready to leave the water together and begin a new life upon the land. This they usually do after a shower, when all the surroundings are moist and at-

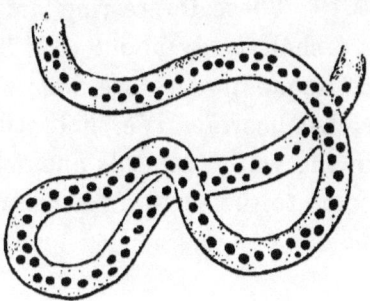

Fig. 140.—Toad's Eggs.

tractive to these dwellers in the marshes, and, from the sudden appearance of the toads, it is a common belief that they have fallen from the clouds with the rain.

15. **The Surinam Toad.** — The Surinam toad has a remarkable way of caring for its young ones. The eggs are laid in the water, and the father at once takes them up and places them on the mother's back, when the skin rises up around them, forming a little cell for each egg. In these curious nests the tadpoles pass through their various changes, remaining here until they become perfect toads.

16. **Tree-toads.** — Tree-toads do not differ much from other toads, except that their toes end in suckers and they can easily climb the trees upon which they live. Most tree-toads are green or brown, and have a general resemblance to the leaves or branches of the trees.

17. **Amphibious Animals.**—Leading this double life, first in the water, then on the land, frogs and toads are called amphibious animals, and you will notice how thoroughly they connect the life history of the fishes with the land animals. They start life with gills and a tail, both of

which they lose, and gain in their places a full set of legs, new lungs, and a third chamber in the heart.

18. These interesting amphibians are the last animals we shall study about that breathe air mixed with water; consequently, we are done with gills, as well as with two-celled hearts. We shall still meet with animals that live in the water, as seals and whales, but they are obliged to come to the surface for their supply of air.

XXXIV.

TURTLES.

SUB-KINGDOM, VERTEBRATA: CLASS, REPTILIA.

1. **An Animal that lives in a Box.** — What a strange idea, that an animal should live inside of its own skeleton, as in a box! But that is just what the turtle does, and you may prove the singular statement by examining the skeleton shown in Fig. 141, or, what is still better, the inside of an actual turtle-shell.

2. **Shell formed of the Backbone and Ribs.** — Here you will find that the upper arch of the shell is made of the flattened vertebræ and the broad ribs, firmly united by notched edges, and held in place by the small pieces of bone near the bottom. There is also a flat bone underneath the body which is not shown in the picture, but which completes the box in which the turtle lives. Openings are left at the front and back through which the animal pokes out its head and tail, and its legs, and when it wishes to, it can draw these parts of the body into the box and shut itself away from the outside world. The shelly plates covering the bone are merely portions of the turtle's skin, hardened into horn or shell.

3. **Box-tortoise.** — The box-tortoise (Fig. 142) is even more thoroughly protected than ordinary turtles. It has joints at the bottom of the shell, so that it can draw up the under parts tightly all around the edge of the box,

10

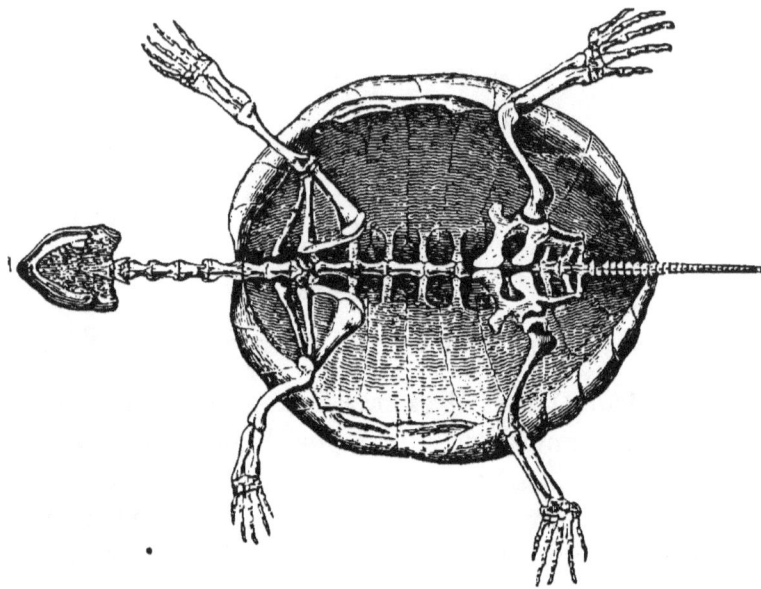

Fig. 141.—SKELETON OF THE TORTOISE (PLASTRON REMOVED).

and when it is thus closed no part of the animal can be reached from the outside.

4. How Turtles Breathe.—You will readily see that the neck and tail are the only movable parts of the spinal column; also, that the ribs, fastened closely together side by side, cannot be used in breathing. On account of this arrangement, turtles are obliged to swallow their air in the same manner as frogs do, but from a different cause. Frogs have no ribs by which the chest is enlarged; and although the turtles have ribs, they

Fig. 142.—BOX-TORTOISE.

cannot use them, because they are immovably fastened to form the shell.

5. **Turtles do not Shed their Shells.**—Some persons have an idea that turtles shed their shells as crabs do. But since we know that the shell is composed of the backbone and the flattened ribs, we may be quite certain that the turtles do not lose these important parts of their body. Turtles have no teeth, but their horny jaws form a kind of beak which serves very well as a substitute.

6. **The Circulation of Cold-blooded Animals.**—The heart of the turtle contains three cavities, two auricles and a ventricle. This formation of the heart gives to all reptiles a peculiar circulation which characterizes them as "cold-blooded animals," and which is of so much interest in this connection that we will try to understand it at once.

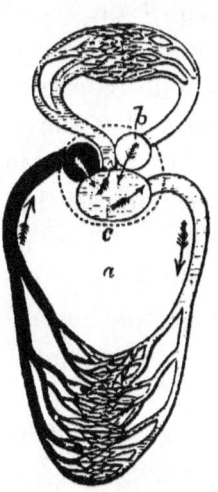

7. After the blood is purified in the lungs, it enters the left auricle (Fig. 143, *b*), while the impure blood from the body enters the right auricle (*a*). The pure blood from the left auricle, and the impure blood from the right auricle, are then both poured into the ventricle (*c*), and this mixed blood is sent all over the system, as well as to the lungs. This imperfect plan of circulation causes the low temperature, the

Fig. 143. — CIRCULA-
TION IN REPTILES.

a, right auricle; *b*, left auricle; *c*, ventricle.

slow breathing, and the sluggish habits of reptiles. Their blood is not well supplied with oxygen, consequently it does not leap and bound through their veins as is the case with more highly developed animals, filling them with energy and activity.

8. Land-turtles and Water-turtles.—Animals of this species which live upon the land generally pass by the name of tortoises; they can swim, when there is occasion for it, but their limbs are better fitted for walking. Turtles and terrapins, on the other hand, live in the water or in marshy places, and their feet form good swimming paddles.

9. Supposed to Live to a Great Age.—It is supposed that tortoises live to be very old, and dates which are sometimes found carved upon their shells might serve to strengthen this impression; but we have no way of judging of the correctness of these dates, and they may not be reliable. Tortoises bearing some peculiar mark are found repeatedly in the same locality, from which we infer that they are not roving in their habits.

10. Hibernating.—None of the reptiles are fond of cold weather, so they do not venture out during the wintertime. These cold-blooded animals have a fashion of hiding away under leaves and brushwood, inside of hollow logs, and in other snug retreats, where they take a good long sleep until pleasant weather comes again, occasionally creeping out on warm days in a stupid condition. This way of passing the winter is called hibernating. Those turtles that live in the ocean go into deep water during the winter, and land-tortoises work their way down into loose, dry earth, where it is quite warm, and where they can sleep undisturbed.

11. Green Turtles.—Green turtles (Fig. 144) are so called from the color of their fat, which is used for food, and which is considered a great delicacy. They live in tropical seas, and are mostly caught when they come on shore to deposit their eggs, although in performing this duty they use the utmost caution. They leave the ocean at night when the tide is at the highest point, and going a short distance from the water's edge, they scoop out a hole in the sand about a foot deep. In this nest they lay a large number of eggs (said to be from eighty to a hundred) and cover them with sand, carefully smoothing the ground to remove any traces of their visit before returning to the sea. All this is done very quickly, and it is almost impossible to tell where the eggs have been placed. Turtles' eggs are round in shape, and are left, without further care, to be hatched by the heat of the sun.

Fig. 144.—GREEN TURTLE.

12. After leaving the egg, the tender young turtles, not more than an inch long, run immediately to the water. In doing so they act prudently, for the sea-birds are fond of them, and eat as many as they can catch. Having reached the water, they are not yet free from danger, but still have many risks to run from fishes and other sea creatures that are ready to enjoy the tempting morsels.

13. **Tortoise-shell.**—Tortoise-shell which is used for ornamental work comes from the hawk's-bill turtle. The plates of this shell are thinner at the edges, as you may see in the picture (Fig. 145), and overlap each other like tiles on a roof. In order to work the tortoise-shell, it is

softened by being placed in boiling water, and while in this pliable condition it readily takes the desired form, which is retained after the shell becomes cold.

Fig. 145.—HAWK'S-BILL TURTLE.

14. When a large piece of tortoise-shell is needed, several plates are joined together. This is done by scraping the edges of the plates very thin and laying them over each other while they are in a softened state. They are then placed under a heavy pressure, and the pieces unite so perfectly that the seam can scarcely be discovered.

XXXV.

SNAKES.

1. **Snakes generally Disliked.**—Lurking as they do in solitary nooks and unfrequented places, and stealthily gliding away when discovered, it is no wonder that snakes have not gained for themselves many friends.

2. Although we know that most snakes are not dangerous, that they are timid animals, anxious to escape from our presence, and, moreover, that their coloring is sometimes rich and beautiful, yet they continue to be objects of general dread and dislike, and few of us can appreciate beauty when it is presented in this repulsive form.

3. **Snakes have no Limbs.**—These animals are generally without limbs of any kind, and we might suppose that in this destitute condition, deprived of the ordinary means of travelling, snakes would be quite helpless; but, on the contrary, they work their way about with great ease, and seem to have no difficulty in catching their prey.

4. **A Flexible Skeleton.**—The skeleton of a snake when carefully prepared, is really a beautiful object. The spinal column is extremely movable, since it consists of a great number of vertebræ, each one working on the next by a ball-and-socket joint. There are sometimes more than four hundred vertebræ, and not only are they joined in this flexible manner, but a pair of ribs is also attached

to each vertebra (excepting a few near the head and tail)
by a ball-and-socket joint, and we can scarcely imagine an
arrangement of bones that would allow more freedom of
motion than this.

5. **The Ribs raised in Creeping.**—There is no breastbone,
and the lower end of the ribs is attached by short muscles to the scales on the abdomen. In the process of creeping the ribs are alternately raised and lowered, so that

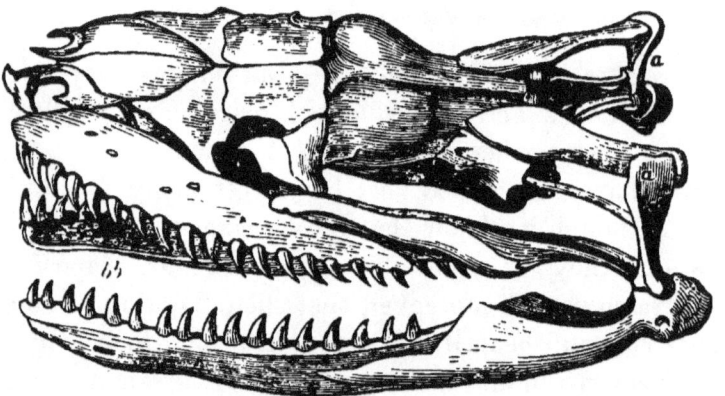

Fig. 146.—SKULL OF BOA-CONSTRICTOR.
a, quadrate bone; b, b, halves of lower jaw.

snakes may be said to walk upon the ends of their ribs.
This movement of the ribs causes the quivering motion
throughout all parts of the body, which you have perhaps
noticed in watching snakes glide noiselessly along, apparently making no effort.

6. **How can a Snake open its Mouth so wide.**—Snakes, we
know, perform wonderful feats in the way of swallowing,
often taking whole animals of greater diameter than their
own bodies. We may be interested in noticing here that
they are enabled to do so by the loose manner in which
the skull is put together, the two halves of the lower jaw

being united in front only by elastic ligaments. In addition to this arrangement, the lower jaw is not joined directly to the skull, but is attached to it by means of a "quadrate bone" between them (Fig. 146, *a*), which is loosely hung, and which allows the mouth to stretch open very wide.

7. **The Teeth.**—The short teeth are pointed backward, and they serve merely to seize and hold the prey. They are not placed in sockets, and therefore are not serviceable for chewing; neither are they needed for this purpose, since the snake swallows its food whole.

8. **The Poison-fangs.** — In most venomous snakes the "poison-fangs" (Fig. 147, *f*) take the place of other teeth in the upper jaw. These poison-fangs are a pair of large teeth perforated by a tube from the poison-gland (*g*). Ordinarily, when not in use, these fangs are laid back and hidden by a fold of the gum, but when the snake is about to strike its prey

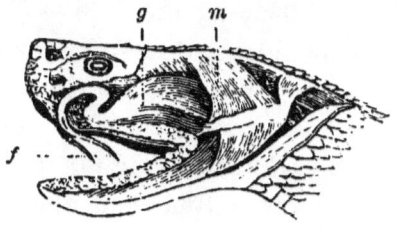

Fig. 147.—Poison Apparatus of the Rattlesnake.

g, poison-gland; *f*, fang; *m*, muscle of jaw.

the fangs spring forward, and the pressure of the muscle (*m*) upon the poison-gland forces the poison down the tube and through the fang ready to flow into the wound.

9. **Snakes coil their Bodies before Springing.** — Before striking with their fangs, snakes generally wind their bodies into a coil, from which they bound like a loosened spring, by straightening themselves out and resting upon the tail.

10. **The Tongue.**—The manner in which snakes stick out their slender, forked tongues looks rather threatening sometimes, but the tongue is perfectly harmless, and need

10*

cause no terror. It is used only as a delicate feeler with which the snake seems to examine objects in passing, frequently drawing it in to moisten it.

11. The Heart shows an incomplete Partition.—The heart of snakes, like that of turtles, has but three cavities, two auricles and one ventricle, still it is of special interest to naturalists on account of an incomplete partition which partly divides the ventricle, and which seems to indicate the probability that after a time we may find animals with perfect four - celled hearts. Let us keep a bright lookout, and perhaps we may soon reach this point in our studies.

12. Shedding the Skin.—The horny scales which form the outside coating of snakes are again covered with a thin, delicate skin which is cast off at certain periods, and a new skin, already formed underneath, takes its place. Before shedding its skin, the animal is quite inactive for a time, and the skin loosens from the body and breaks around the mouth. As the snake then creeps through some crevice or close brushwood the skin is drawn off inside out, much in the same way that we sometimes draw a glove from the finger.

13. The Staring Eyes.—The fixed, stony stare which is so unpleasant a feature of snakes is due to the fact that they have no eyelids. Many thrilling stories are told of snakes having charmed birds and other small animals, but as the stories do not seem to be well authenticated, there is little reason to suppose that snakes possess any such power. The birds may, in these cases, have been overcome by a sense of fear, for the snakes are their most deadly foes.

14. The Rattlesnake. — None of our snakes are more dreaded than the rattlesnake, whose bite is exceedingly poisonous. This snake generally gives warning, however, before striking its prey, by shaking a peculiar rattle at the end of its tail. The rattle is composed of a number of

Fig. 148.—RATTLESNAKE.

horny, button-like rings which fit loosely into one another and make a rustling noise when shaken rapidly. It is believed that one new rattle is gained with each shedding of the skin, but as those at the end sometimes fall off, the age of the snake cannot be told with any degree of certainty by the number of rattles. Young rattlesnakes are hatched in broods of eight or ten early in the summer, and keep together until they are pretty well grown. In times of danger they are said to run into the mouth of their mother for protection, the same as other young snakes do.

15. **Boas.**—The tropical swamps of South America form a congenial home for serpents, and here are found the great boas, which are the largest of living snakes, sometimes exceeding twenty feet in length. The bite of the boa is not poisonous, but these animals are dreaded on account of their great muscular power. They coil themselves tightly around their victims, and in this way destroy life in a few minutes, even breaking the bones.

16. **Manner of Attacking their Victims.**—Boas have curious hooks or claws near the tail which are in reality the traces of hind limbs. By these hooks they hang suspended from the branches of trees, aided also by twining the powerful tail around the branches. Waiting in this favorable position near a stream until some unfortunate animal comes to drink, they coil themselves with wonderful rapidity around their prey and crush it into a shapeless mass, fit to be swallowed. After a full repast they usually lie in a stupor for some time before waking up to need more food. While in this partially unconscious condition the boas themselves are subject to many perils from their enemies, who well know the effect upon these greedy creatures of over-indulgence of the appetite.

17. **The Cobra.**—The cobra of India is an exceedingly venomous snake. It is a small snake, only three or four feet long, with a curious arrangement by which the skin of the neck may be made to cover the head like a hood,

when this dreaded creature is irritated. On this account it is sometimes called the hooded snake, and it is the one usually carried about by snake-charmers. The cobra is very sluggish in its habits, and it has an unpleasant fancy

Fig. 149.—COBRA DE CAPELLO.

for entering dwelling-houses, which, in this hot country, are infested by many unwelcome visitors. Every year in India several thousand people die from the bite of the cobra, and no cure has yet been discovered for the deadly poison, although the British Government has offered a tempting reward for such a cure.

XXXVI.

LIZARDS.

SUB-KINGDOM, VERTEBRATA   CLASS, REPTILIA.

1. **Accomplished Climbers.**—The active lizards with long, slender tails differ from snakes principally in having four well-developed limbs, each ending in five toes; and these they use very nimbly in running over rocks and along fence-rails. The position of their own bodies seems to make little difference to the accomplished climbers, and they have no difficulty in running upon smooth under-surfaces with their backs downward.

2. **The Eyes are protected by Lids.**—These bright-eyed little creatures have two eyelids which they draw over the eye in winking, and as they have none of that stony expression so disagreeable in the snake, we do not mind their gazing at us. In fact, it is quite entertaining to gaze at them in return and see how rapidly they dart out their slender, forked tongues to catch the insects which form their chief food.

3. **Lizards' Tails are very Brittle.**—One odd thing about lizards is that their tails are so brittle, snapping off sometimes very unexpectedly. Such being the case, there is no security in holding a lizard by its tail, for the cunning little creature may run off and leave both you and the tail you are holding. This is no serious loss to the lizard, however, for a new tail soon grows in its place.

4. **Places frequented by Lizards.**—Some kinds of lizards love to frequent sandy banks where they may bask in the sunshine, while others conceal themselves in damp places under stones and rotten logs. Fortunately for the shy and timid lizards, their metallic colors often resemble the vegetation or the soil on which they live, and they can thus easily keep out of sight.

Fig. 150.—LIZARD.

5. **Advantages gained by Mimicry.**—With regard to this similarity of coloring, or mimicry, so often adopted by animals, it is important we should understand that it is a protection which these animals have gained or cultivated for themselves—unconsciously, of course, and entirely without effort on their part, but as a natural result of the fact that animals which are so protected stand a better chance for life. Those animals of any kind which

are inconspicuous among their ordinary surroundings are the ones most likely to escape their enemies, and at the same time they are the ones to grow strong and healthy from having plenty to eat, since it is easier for them to pursue their prey without being seen. This advantage once gained is held on to, and, having proved useful to the parents, is handed down to their offspring, thus increasing little by little, and producing at last very curious and interesting results.

6. **The Chameleon.** — Many lizards have the singular power of changing their color, which gives them anoth-

er advantage in concealing themselves. The chameleon, especially, is celebrated for its sudden changes of color (Fig. 151). It lives only in the warm parts of the Old World. It is extremely dull

Fig. 151.—The Chameleon.

and sluggish, showing no spirit for anything but catching insects; and at this favorite sport it is quick enough. Sometimes it remains for hours in one position, making no movements except as some insect goes buzzing past, when the sticky tongue is instantly darted out to catch it. This tongue, so destructive to insect life, is covered with sticky saliva, and looks as if it were swollen at the end. It is fastened at the front of the lower jaw, and is usually coiled up in the mouth, but when thrown out it may be extended for a length of six inches.

7. As chameleons pass their time in the trees, they need to hold on to the branches, and for this purpose their toes are arranged in two groups, like a parrot's, three toes on one side of the foot, and two on the other. This arrangement enables them to grasp the twigs with their toes, and at the same time they twist their tails around the branch, as you see in the picture. Any part of an animal which is fitted to seizing or grasping objects in this way is said to be "prehensile." Thus, we speak of the prehensile tail of the chameleon. With this definition, what do you think we might now say of the tail of the boa?

Fig. 152.—THE IGUANA.

8. **The Iguana.**—The iguana of tropical America is a very large lizard, nearly five feet long, with a row of bristling points standing upright on the middle of the back and tail. The long, awkward legs end in sharp claws, which assist in climbing trees, and, clumsy as this creature looks to be, it runs about among the branches with great activity.

9. **The Flying Dragon.** — Another curious lizard, much more attractive than the iguana, is the flying dragon of

Fig. 153.—THE HORNED TOAD.

the East Indies. It has a broad, wing-like fold of skin on each side of the body, supported by two ribs which stand out straight from the spinal column. The dragon is upheld by this skin as it takes flying leaps from one tree to another, but it has no power of striking the air as a bird does with its wings.

10. **The Horned Toad.**—The horned toad, so characteristic of our Western plains, is in reality a lizard. Its young ones do not pass through the curious changes which all young toads pass through; it has a real, honest tail, and it runs like a lizard instead of hopping as toads do. Persons visiting the West often bring home these rough-skinned horned toads for pets, and we hear of their living for months without food; but during this time they may perhaps have snapped up a goodly supply of flies when no one was looking.

XXXVII.

CROCODILES.

SUB-KINGDOM, VERTEBRATA : CLASS, REPTILIA.

1. Crocodiles in their Native Homes.—To see crocodiles in perfection we should visit the rivers of tropical Asia and Africa. Here these huge bronze-green reptiles sometimes reach the length of thirty feet, and being strong and ferocious, they are really dangerous animals.

2. It is difficult for us to understand how these repulsive creatures ever came to be regarded with favor by the natives, who must often have suffered from their attacks, but, strange as it may appear, they were considered sacred by the ancient Egyptians, and were trained by them to take part in their religious processions.

3. The Ferocious Mouth. — The mouth of the crocodile looks particularly ferocious, as there are no lips to cover the grinning, pointed teeth, which are always visible. These are real, strong, biting teeth, too, and they are firmly set in pits in the jaw, and below the root of each tooth there is a little new tooth started, ready to grow up and take the place of the old one if it should fall out. In this way vacancies are soon filled, and no matter how old a crocodile may be, its mouth is always supplied with a full set of teeth. As if to add to this hideous effect, the fourth tooth in the lower jaw is longer than its neighbors, and when the mouth is closed it extends up over the upper jaw.

Fig. 154.—CROCODILE-HUNTING.

4. **The Tough Hide.**—The crocodile's hide is exceedingly tough, composed of plates of bone covered with horny scales. These plates are raised into ridges on the back and tail, where they form various patterns.

5. **How Crocodiles capture their Prey.**—Their powerful tails, besides being useful in swimming, are very convenient for striking down and sweeping into the water many of the large animals upon which they feed. Having thus secured their prey, they hold it under water until it is drowned, being careful, however, to raise their own snouts

above the surface once in a while for air. The mouth of the crocodile is necessarily open during this time, but by a peculiar arrangement of valves, shown in Fig. 155, the throat and nostrils may be instantly closed, so that water cannot run down the throat; and the process is perfectly safe for the crocodile, although disastrous to its victims. The wily crocodile then drags the lifeless body ashore, often hiding it until it is partly decayed before eating it.

6. **Their Difficulty in Turning.** — It is generally known that crocodiles have great difficulty in turning quickly, and when persons are pursued by them they very frequently profit by this knowledge and make their escape by rapidly changing their course. The want of flexibility noticed in such cases is caused by the small ribs which are attached to the vertebræ of the neck, and which interfere with its free movements.

Fig. 155.—MOUTH OF THE CROCODILE.

d, tongue; e, glands; f, inferior, and g, superior, valves separating the cavity of the mouth from the throat, h.

7. **The Heart and the Circulation of the Blood.** — We must not fail to examine the heart of the crocodile, for we shall find in it a decided step in advance of the other reptiles. The change which we had reason to expect has at last taken place. You will remember that we noticed an incomplete partition in the ventricle of the snake, but

Fig. 156.—ALLIGATOR.

here there is a perfect partition making two separate ventricles, and the heart is thus divided into four distinct cells. The pure blood is consequently kept separate from the impure blood as long as it remains in the heart.

8. The advantage gained by this arrangement is soon lost, however, for the main arteries carrying both the pure and the impure blood unite soon after leaving the heart,

and the two kinds of blood flow together in one stream. A mixed blood is therefore sent through the body of the crocodile, the same as in other cold-blooded animals. Notwithstanding this apparent failure, a perfect circulation, by which the system is supplied only with pure blood, is nearly reached, as we shall find when we come to treat of birds.

9. **Alligators.** — Alligators are very much like crocodiles, except that they are smaller, that their feet are not so completely webbed, and that they are found only in America. They are very numerous in the rivers and swamps of our Southern States, where they may be seen swimming with the snout just above water, or sunning themselves on the banks; and they are a never-ending source of entertainment to visitors and sportsmen in these popular winter resorts. These animals are most active during the night, at which times they make a loud, bellowing noise. In cold weather they bury themselves in the mud and become perfectly torpid, but the warm sunshine soon revives them.

10. **The Eggs of Alligators.**—Alligators lay a large number of eggs, which are deposited on the river-banks in a mass of vegetable matter heaped up for the purpose, and the decaying of this mass produces the heat necessary for hatching the eggs. The mother keeps watch over the spot, and tears open the pile

Fig. 157.—JUST HATCHED.

to liberate the young when she hears their cries. She shows some affection for them, and does what she can to protect them from the old males, who will eat them if they have a chance to do so.

11. **The Reptiles of Ancient Times.** — All kinds of reptiles are most numerous in warm countries, and they have many points of resemblance to birds. There was a time, very long ago, when reptiles of all sizes must have been abundant on our earth. Some were small, like the lizards of the present day, while others were of gigantic size, measuring fifty or sixty feet in length. We know of

Fig. 158.—FOOTPRINTS OF LABYRINTHODON.

these odd reptiles only by the hints and sketches of their lives which are traced in a most convincing manner upon the rocks in the form of fossil remains, and many of them are exceedingly interesting because they are so different from any animals now living.

12. Some of these old-fashioned creatures had long necks, and wings which in all probability were used by them for mounting into the air, yet at the same time the upper-jaw contained many teeth set in distinct sockets like those of a crocodile, and, upon the whole, the skeleton would suggest a bird quite as much as a lizard. The oldest of these fossils yet discovered which naturalists can place confidently with birds is the *archæopteryx*, and a strange-looking bird this one must have been. It had a long, jointed tail like a lizard's tail, with one pair of quill feathers standing out from each vertebra.

13. Then there were other forms entirely different from these birdlike reptiles — huge beasts having very long hind-legs, upon which they probably hopped or walked nearly erect, while their short front legs could scarcely have touched the ground. Others, again, had strong tails and paddle-feet, suggesting at once the thought that they must have been good swimmers.

14. The rocks in the Connecticut Valley are marked with a great variety of birdlike tracks, some of them very large, which are now believed to be fossil footprints of curious reptiles like some of those we have been speaking of.

15. In Fig. 158 are shown fossil footprints of the labyrinthodon, an animal which was first known to science only by its footprints; and these impressions bear so strong a resemblance to the human hand that the animal has been called the "hand-beast."

11

16. In looking at these footprints, we can form no idea of the vast number of years which must have passed since these old-time creatures chanced to step on this particular spot of mud, leaving impressions to be treasured up long after their race had died out, and the earth had become peopled by much more attractive beings.

XXXVIII.

CHARACTERISTICS OF BIRDS.

SUB-KINGDOM, VERTEBRATA : CLASS, AVES.

1. **The Beautiful Birds.**—There is something very winning about birds, and perhaps we are scarcely conscious of how much these light-hearted creatures do to gladden our lives. Joyous and beautiful, they charm us with their graceful movements, no matter whether we see them soaring through the heavens, flitting about among the tree-tops, or chirping contentedly over their simple every-day duties. Then their sweet and varied song is another source of delight, to which few persons are insensible.

2. **What makes Birds so Light?**—Most birds, we know, can leave the earth and mount up into the air, passing the greater part of their lives in regions that other animals are entirely unable to reach. We may safely conclude that their bodies are constructed differently from other animals or they could not thus enjoy so unusual an advantage. Upon looking closely into the matter we shall discover that birds are made as light in every way as possible. Their bones are hollow, the quills of their feathers are hollow, and both of these are filled with air; moreover, the feathers themselves are made up of delicate filaments which cling to one another and hold a great deal of air in their meshes.

3. The drawing of the vulture in Fig. 160 will give us

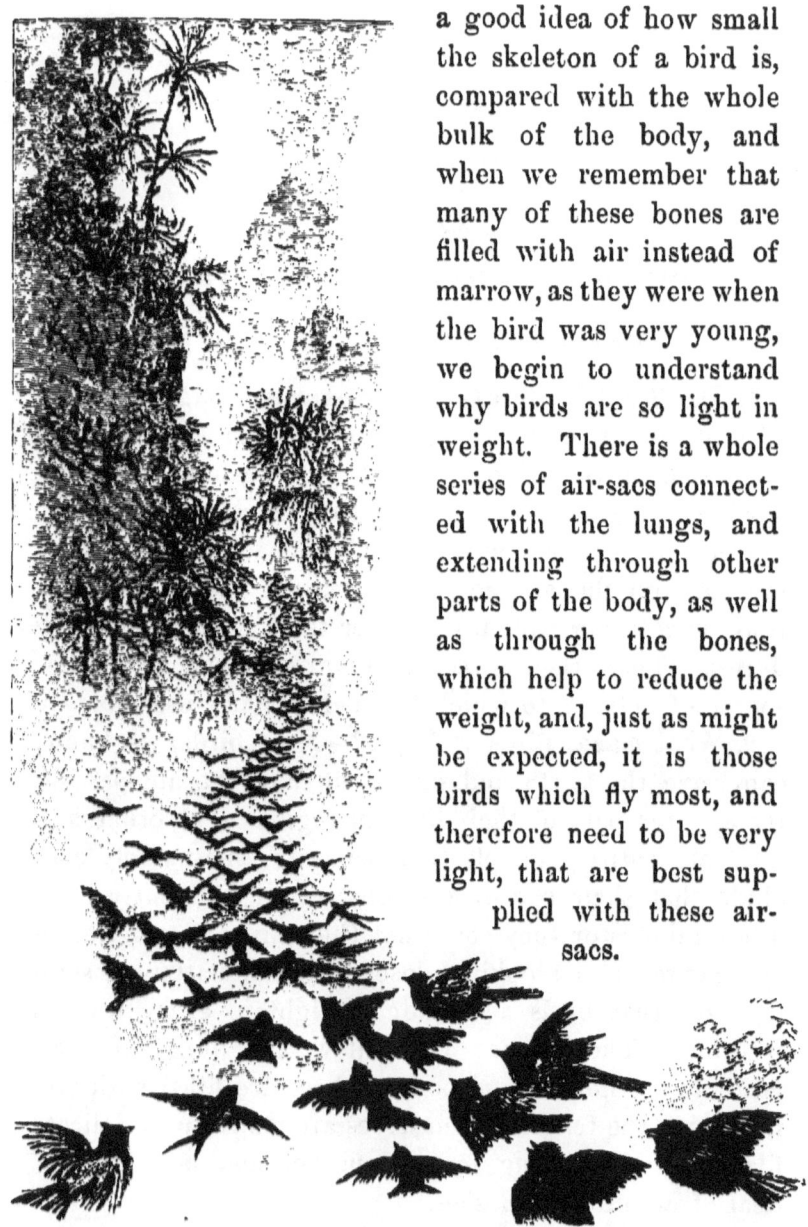

a good idea of how small the skeleton of a bird is, compared with the whole bulk of the body, and when we remember that many of these bones are filled with air instead of marrow, as they were when the bird was very young, we begin to understand why birds are so light in weight. There is a whole series of air-sacs connected with the lungs, and extending through other parts of the body, as well as through the bones, which help to reduce the weight, and, just as might be expected, it is those birds which fly most, and therefore need to be very light, that are best supplied with these air-sacs.

Fig. 159.—"The North-wind."

4. The Wing of a Bird.—Still, for all this, the bird could not fly without wings. So it is the wing that charms us most, and when we see what a simple thing it is, we wonder at its power. The framework is formed of a set of bones (Fig. 161) very similar to those of our arm and hand, but having only one perfect finger, which corresponds to our index-finger. Stretched over this framework there is a thin covering of flesh and muscle from which grow the quills and smaller feathers, and these, when spread out, make up the broad wing.

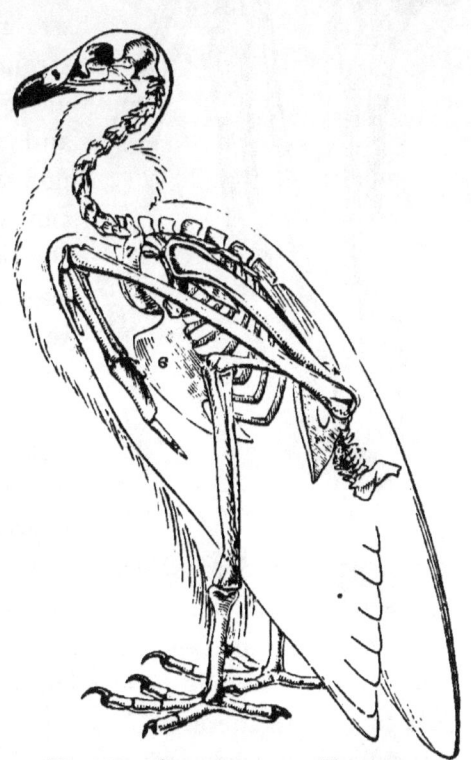

Fig. 160.—Skeleton of a Vulture.
6, keel of sternum; 7, clavicle, or wish-bone.

5. When opened, the upper surface of the wing is rounded, and the air can readily slide from its edges as the bird mounts upward, while, on the contrary, the hollowed under surface holds the air, as in an inverted cup, and enables. the bird to press upon the air thus confined. The movement of the wing in flying is somewhat like the stroke of an oar; the wing cuts the air with its sharp front edge, but presses back upon it with its full outstretched surface. As this action is repeated again and again,

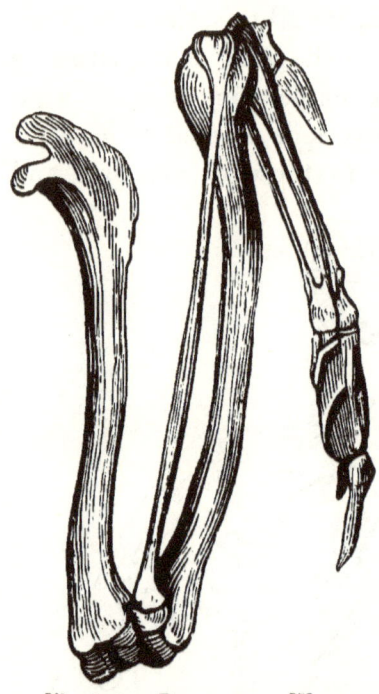

Fig. 161.—Bones of a Wing.

the bird moves forward with an easy, gliding motion.

6. **The Flying Muscles on the Breast.**—Although flying looks to us like easy work, we may know there are strong muscles required to move the wings so rapidly and so gracefully. These powerful muscles that move the wings are situated on the breast, and they are fastened at one end to the large breastbone or sternum. In those birds that fly, the breastbone extends straight out in front like the keel of a ship (Fig. 160, 6) and gives a good surface for the attachment of the muscles. It is these same flying muscles which make the breasts of birds so full and plump.

7. **The Wish-bone.**—The wish-bone (Fig. 160, 7) which delights the heart of every child has also an important part to play in the act of flying. It is formed of the two collar-bones united in front, and its particular office is to keep the shoulders apart and prevent the wings from sliding towards the breast while the bird makes its downward strokes. This little wish-bone is another good example of the effect of use or disuse upon certain parts of the body. In ostriches and other birds which do not use their wings for flying the bone is not needed; consequently, it does not grow, but always remains in an undeveloped condition.

8. **The Arrangement for Perching.**—Have you ever wondered how it is that a little bird can sit so securely upon its perch all night long and never once tumble off when it begins to ·nod and finally goes fast to sleep? In Fig. 162 we can see what a beautiful arrangement there is for perching. The thick muscle which bends the bird's toes is placed above the knee; the fibres composing this muscle unite into a long, white cord, or tendon, which passes down in front of the knee (Fig. 162, A), then winds around to the back of the heel and goes to the toes.

9. As the bird settles on the perch and bends its knee the weight of the body pulls this tendon and involuntarily bends the toes. For this reason the toes clasp the perch securely even when the bird is not thinking about it. The weight of the body continually pulls on the tendon, and the bird cannot help holding fast to the perch. In the same way, when a bird is walking it cannot keep its foot open while its leg is bent.

Fig. 162. — LEG OF A BIRD PERCHING.

This peculiarity you have no doubt noticed in a chicken stepping slowly, with one foot daintily uplifted and seeming to scorn the idea of again touching it to the earth.

10. **The Food of Birds has an Important Bearing upon their Structure.** — By watching birds we shall discover that, while some feed upon insects, other kinds are hunting for fruits and seeds, and this question of food has an important bearing upon their habits and choice of a home, and also upon the structure of their bodies.

11. Birds which feed on easily digested animal food have a simple gizzard with thin walls. On the other hand, those birds which feed on hard grains have thick, muscular gizzards for grinding their food. The little

pebbles which birds swallow are necessary to help in this grinding, and without them the gizzard could not crush the food properly.

12. The digestive organs of a bird, as seen in Fig. 163, may be easily studied in the common fowl. The long œsophagus (3) is enlarged to form a crop (4). Here the food is moistened with digestive juices, and then passed on to the gizzard (7) in small quantities.

13. **A Perfect Four - celled Heart.**—After noticing the gradual modifications in the hearts of fishes and reptiles, we have now the satisfaction of finding in birds the first instance of a perfect - working four - celled heart, which succeeds admirably in keeping the pure blood and the impure blood from becoming mixed.

14. The right side of the heart, as shown in the ideal plan in Fig. 164, is set apart to receive the impure blood and no other, and to send it to the lungs, while the left side receives the pure blood that is returned from the lungs and

Fig. 163.—DIGESTIVE ORGANS OF A FOWL.

3, œsophagus; 4, crop; 7, gizzard.

pumps it into the arteries, which carry it to all parts of the body.

15. **The High Temperature of Birds.**—This perfect circulation is found in all the higher animals, giving them a high temperature. Still, the blood of birds is hotter than that of other animals, owing partly to their perfect circulation, to the abundant supply of air in their bodies, and also to their covering of down and feathers, which keeps the heat from escaping. The temperature of birds is 104°; that of the human body is 98°.

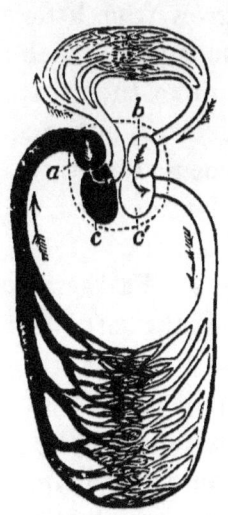

16. **The Music of Birds.**—The exquisite music of the birds is produced by an arrangement of bones and cartilages at the lower end of the windpipe, or trachea, as it is called. Currents of air passing through this part of the trachea give a quivering motion to the membrane stretched across it, and produce all the varied notes so pleasing to our ears.

Fig. 164.—PLAN OF CIRCULATION IN BIRDS AND MAMMALS.

a, right auricle; *b*, left auricle; *c, c'*, ventricle.

17. Birds evidently sing sometimes for their own pleasure, and then the whole depths of their nature seem to be poured into their music; but their song is also at times a call to their mates and to their young ones. As a rule, male birds are more musical than their mates. They are likewise larger and more brilliant.

18. **Their Coloring.**—Gayest of all the birds are those living in the tropics, while those that inhabit the Arctic regions are of dull colors, and some of them change to white when winter comes and the ground is covered with snow. So birds, too, you see, also resort to mimicry.

11*

19. **Their Plumage.**—Birds are the only animals that are clothed with feathers, and much of their beauty is due to the colors and markings of this soft, fluffy covering as well as to its charming metallic lustre.

20. **Growth of the Feathers.**—These beautiful feathers grow from little sacs in the skin, and are made of a horny substance, much the same as the scales of reptiles. But, unlike these scales, the feathers have split up, during growth, into many narrow strips, which give them their peculiar softness. Upon the legs and feet of most birds you may see scales which have not split up thus into feathers.

21. **Feathers examined.**—You may also notice how different the soft, downy feathers, overlapping one another so as to form a warm covering for the body, are from the large quill-feathers of the tail and wings. These last are very useful in flying, since the tail feathers form a kind of fan which helps the bird to steer its course, and the wing feathers greatly increase the size of the wing without adding much to its weight.

22. Upon examining any common feather you may see on each side of the quill small branches, or "barbs," and you must notice how completely these unite with each other to form the broad "vane" (c, Fig. 165). On both edges of the barbs there is a row of still smaller "barbules," ending in hooks, which interlock with the next row and are thus held firmly together. When these barbs are pulled apart we can see the little barbules separating, and if they are placed together again side by side they will unite as before.

23. Now, we cannot make the lower barbs on the quill unite in this way, neither can we the down feathers which lie next to the skin of the bird, because their barbules have

no hooks on them. This is also the case with the feathers of the ostrich and other birds of its kind—the barbs never unite with each other, and hence the plumes are soft and downy.

24. **Preening.**—The neck of birds is always long enough to allow the beak to reach an oil-gland which is situated at the end of the tail, and which supplies the oil for "preening" the feathers. The tail can also be raised part way to meet the beak, and the bird, having thus obtained a supply of oil from the gland, passes its feathers one after another through its bill in such a way as to distribute the oil through the plumage. This process of preening smooths the feathers so that the bird may glide easily through the air or water, as the case may be, and it also serves to make the plumage water-proof.

25. **The Migration of Birds.**—But few birds remain constantly in the places where they are hatched. Migratory birds unite in flocks and take long journeys at certain seasons, leaving cold countries at the approach of winter and returning in the spring, thus making two journeys each year. It is an interesting fact that they always make their nests and raise their young broods in the coldest countries which they visit.

Fig. 165.—Parts of a
Feather.

a, quill; b, shaft; c, vane;
d, down.

26. In these migrations all the birds of one species seem filled with an impulse to move in one direction in search of food or other favorable conditions. Each kind has its time for starting, and seldom varies from it. They sometimes return the following season to the exact spot they started from, having in the mean time travelled hundreds of miles. The sea does not stop them, but they often take long flights over its surface, as, for instance, in crossing the Mediterranean Sea from Europe to Africa, or in going from our own coast to the Bahama Islands and to South America.

27. **Birds Useful in destroying Insects.** — Many of the smaller birds, which in former times were killed or frightened off because they robbed our gardens and orchards, are now considered useful in destroying insects, and gardeners are doing what they can to invite their return. That birds eat fruit is very true, and that they select the largest and finest is also true; nevertheless, the injury done by insects is far more serious in its character.

XXXIX.

BIRDS' EGGS AND NESTS.

1. **The Egg.**—What a mystery is connected with the egg! A little world of itself! Shut apart from the outside world, it seems a lifeless thing, yet within that little sphere mighty forces are at work, which, under favorable circumstances, will produce a perfect animal, gifted with life, and soon showing the habits and peculiarities of its ancestors.

2. **The Study of the Egg.**—On opening an egg we see merely the "white," in the middle of which floats the "yolk," with the whitish "germ cell" clinging to it. This germ cell occupies but little space, yet it is the important part of the egg—the part for which all the rest of the egg was made, because it is just at this spot that the young bird begins to grow. We cannot see without a microscope the twisted cords of albumen at both ends of the egg which hold the yolk pretty nearly in the centre, but we can see them represented in Fig. 166. Those twisted cords allow the yolk to roll over from one side to another when the egg is turned, and so the germ cell, which is at the lightest part of the yolk, keeps always uppermost, as in the picture. Here we have a beautiful contrivance by which the germ cell is sure to be nearest to the body of the bird as she sits upon her eggs, no matter how often the eggs are turned over.

3. **Why a Bird sits upon her Eggs.**—Of course, that part of the egg nearest the bird gets the most heat from her warm little body and her soft, downy feathers, and a certain amount of heat is necessary to develop the new life within the egg. This we know is the reason that birds sit upon their eggs, and that they are so careful not to leave the nest long enough for them to become chilled.

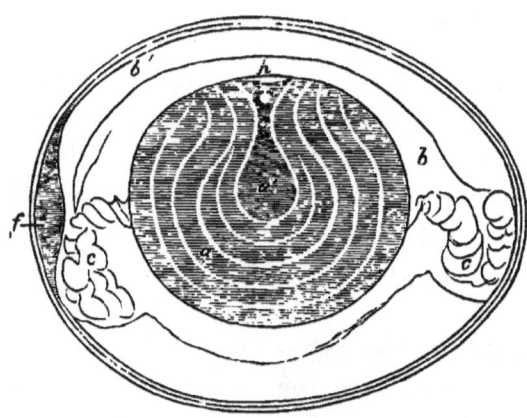

Fig. 166.—SECTION OF A HEN'S EGG BEFORE INCUBATION.

a, yolk, showing concentric layers; *a'*, its semi-fluid centre; *b*, inner dense part of the albumen; *b'*, outer thinner part; *c*, twisted cords of albumen; *h*, the white spot, or germ cell.

4. **Growth of the Young Bird.**—As we have just said, the young bird begins to grow from the germ cell. The albuminous white of the egg furnishes the building material for its growth, and the rich oily yolk nourishes the newly formed bird as long as it continues in the shell. The more there is of this rich yolk stored up in the egg, the stronger and better developed will the bird be on leaving it, as is clearly shown in the case of those birds whose eggs contain a large yolk. The young of such birds are able to run about and help themselves as soon as they are hatched; whereas the young of those

having small yolks, not being so fully developed, are hatched in a blind and naked condition, and need to be fed and brooded over by their parents.

5. **The Supply of Air.**—No doubt you have often noticed in hard-boiled eggs a little hollow place at the larger end like the one shown at *f*, in Fig. 166. There is a little bubble of air here, between the two deli-cate tissues lining the shell, for the use of the

Fig. 167.—Building the Nest.

baby bird, and the shell is also full of small pores through which fresh supplies of air can easily pass.

6. **The Bird chips its Way out.**—When the tiny creature, shut up in the shell, is fitted to live in the great world outside, it pierces this hard case and chips its way out by the help of a small knob on top of its beak. This knob seems to be only a tool to help the bird escape from the shell, and as it is of no use afterwards it soon disappears.

7. The bird is now fully equipped with bones, muscles, bill, claws, and internal organs. These parts have all been formed and nourished from the contents of that little eggshell. Moreover, we find the contents of the shell have been entirely absorbed, showing that though the egg furnishes all that is needed for the formation of the young animal, there is nothing in it which is unnecessary.

8. **Birds' Nests.**—These same birds' eggs, so full of wonderful design, are very precious to the heart of the mother-bird, and she never seems happier than when hard at work getting her nest ready to receive them. The nest is also intended for the early home of her little ones, and she displays much skill and industry in building it. As a general rule, the small birds with delicate feet and slender bills are most successful in weaving a fine and elegant nest.

9. All birds of the same species build their nests alike from one generation to another, and seldom depart from the long-established plan. They not only use the same building material, but they select similar locations, so that those who are familiar with the habits of birds know pretty well in what kind of places to look for any particular nests they may be in search of.

10. Birds' tastes differ widely in the choice of a home,

and high tree-tops, way-side hedges, low bushes, hollow tree-trunks, and grassy pastures all have advantages of their own in the estimation of the birds that occupy them. Every boy and girl should know the keen pleasure of finding these charming nests hidden away among the leaves and grass.

11. **Building Materials.**—Some birds, you may have noticed, use nothing for building materials but small sticks, dried grass, and hair; some weave pieces of string and strips of birch bark in among the grass; others, again, plaster their nests with mud to make them strong. The great-crested flycatcher has a singular fancy for the cast-off skins of snakes, and always hunts up one or two of these skins to weave into her nest. She then lines it with soft brown feathers of the same general color as the eggs that are to lie within it. The tailor-bird also makes an odd nest (Fig. 168) by sewing together the leaves of trees, and in doing so she must use her beak and slender claws in the place of a nee-

Fig. 168.—Nest of the Tailor-bird.

dle. In arranging their nests most birds have a thought for comfort, and put in some soft lining, using for this purpose feathers, fine grass, delicate thistle-down, or the yellow woolly covering of young ferns.

12. **Dangers to which the Nests are exposed.**—Imagine, now, these dainty homes after the tiny eggs have been placed within, or a little later, when they are filled to overflowing with tender young birds, and you may know how attractive they are to hawks and owls and snakes

and other animals that are prowling about, seeking what they may devour. In fact, these little tidbits are so eagerly sought both day and night as to make the parent birds very anxious for the safety of their little ones, and

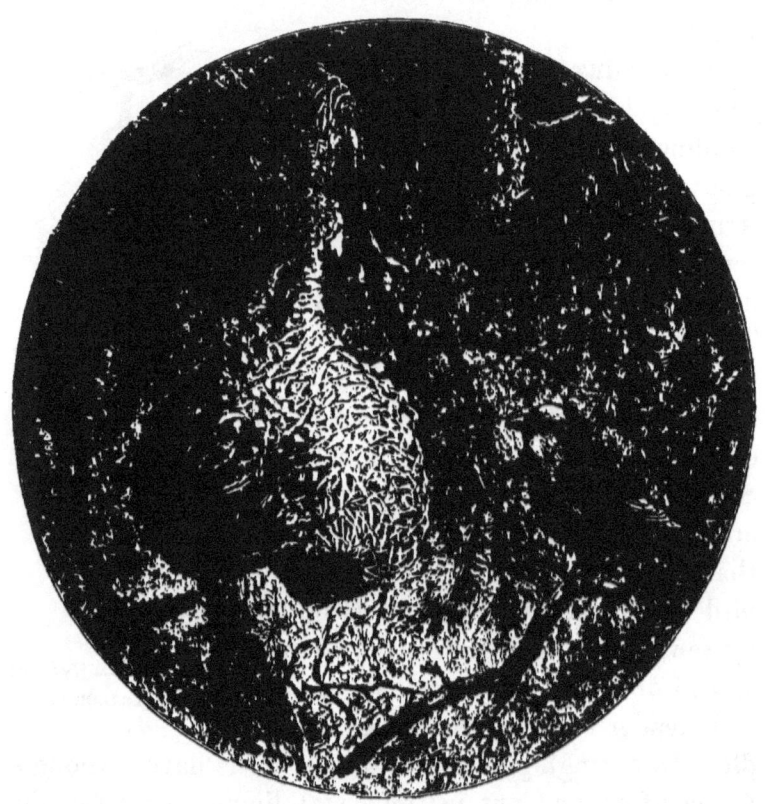

Fig. 169.—Nest of the Weaver-bird.

in consequence of the dangers to which they are exposed most birds conceal their nests as much as possible from sight. In the tropics they often hang them on the outer twigs of trees, away from the reach of monkeys and reptiles. The illustration on this page shows the nest of the African weaver-bird, curiously fashioned, and hanging from the branch of a tree.

13. **The Necessity for Screening a Bright-colored Bird.**— It has been noticed that those female birds which have bright and conspicuous colors, like their mates, build in hollow trees, or else make covered nests, that they may not be so easily seen while sitting upon them. On the other hand, when the female is of a dull color, and there is not the same need of concealment, the nest is made open. It will at once be evident that a bird which harmonizes in color with the general hue of her nest might sit upon it unnoticed, whereas a bright - colored bird in such an exposed position would attract the attention of her enemies, and thus inform them where her treasures are stored.

14. **Eggs concealed by their Coloring and Markings.**— This fascinating subject of the coloring of birds may be extended to the eggs as well, and you will find it a pleasing study to notice the various tints by which birds' eggs are made to blend with their surroundings. The curious blotches and specks and the indescribable lines and markings with which many eggs are ornamented serve as an additional concealment. Perhaps you will discover that eggs which are placed in open nests are generally shielded from observation in this way, while those eggs that are laid in holes and in concealed places are often purely white.

XL.

SWIMMING–BIRDS (*NATATORES*).

SUB-KINGDOM, VERTEBRATA : CLASS, AVES.

1. **Birds divided into Seven Groups.**—Nowhere in the study of Natural History do we find animals more beautifully fitted to the kind of life they lead than the birds are. With regard to their habits and manner of living, they seem naturally to divide themselves into the following seven groups: the *swimmers*, the *waders*, the *runners*,

Fig. 170.—WILD DUCKS.

the *scratchers*, the *climbers*, the *perchers*, and the *birds of prey*. Perhaps we cannot do better than accept the hints of nature, so we will study them in this order. The beaks and feet, as a general thing, indicate the habits of the bird, and show to what group it belongs.

2. **Ducks and Geese: Examples of Swimming-birds.**— Ducks and geese present themselves at once to our minds as familiar examples of swimming-birds, and we can see how exactly their boat-shaped bodies and long necks are suited to living on the water. Then they are web-footed

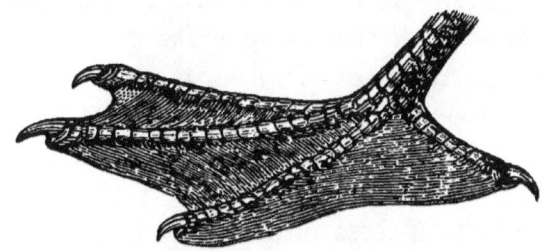

Fig. 171.—FOOT OF A GANNET.

(that is, they have a skin between the toes, as shown in Fig. 171), which enables them to swim easily, and their short legs are placed far back on the body. This position suits very well for paddling in the water, but it makes their gait upon land extremely awkward—so much so that the waddling of the duck has become proverbial.

3. **Protection against Changes of Temperature.**— First diving into the water, then flying up into the air, these swimming-birds are necessarily exposed to great changes of temperature, and as a protection against such sudden changes their bodies are covered with a thicker and closer plumage than other birds have ; the coat of down next to the skin is also very thick. There is, moreover, an un- usually large supply of oil from the oil-gland, which keeps

the plumage from getting wet, and gives the feathers that beautiful gloss so noticeable in the duck.

4. The Bills form a good Strainer.—These birds have an odd way of gobbling up their food, often taking in mud and water at the same time, but their broad, flat bills are furnished with rough plates around the edge, which form a very good strainer. Being richly supplied with nerves, this strainer is able, in some way, to select the particles of food and keep them in the mouth while the mud is allowed to run out.

5. The Swans.—More graceful than the ducks and geese are the swans, with their long necks gently curved and their wings partly lifted when swimming, as if to catch the wind. When swans are attacked, they defend themselves with spirit, making a loud hissing sound and striking violently with their wings.

6. How Flocks of these Birds may be Recognized.—Wild ducks, geese, and swans are excellent flyers as well as swimmers, and they can be recognized at a distance by their wedge-shaped flocks. In reality, these birds fly in two lines, which come together in front and gradually separate towards the last of the flock, so that the general appearance of the company has the shape of a wedge. The leader flies at the point where the two lines meet, and when he becomes weary, he leaves his post to his next neighbor, and falls back to the last of one of the two lines; but meanwhile, during this change of leaders, the flock keeps in perfect order. Upon these migrations the birds fly thousands of miles, and build their nests in summer among the lakes and marshes of cold northern countries.

7. Flamingoes and their Nests.—A company of flamingoes like those in Fig. 172, with their bright scarlet bodies, their long slender legs, and their curved bills, is certainly very handsome; but how odd they must look when sitting on their high, conical nests, which are made of mud heaped up into slender mounds two or three feet high, and shaped somewhat like an old-fashioned churn—small at the top, and increasing in size towards the bottom. These mounds are scooped out at the top, just enough to hold the eggs, and the bird sits upon the column with its legs doubled under it, in the same way as other birds sit upon their nests.

8. **Sea-birds.**—Less familiar than some of these land-birds, but not less interesting, is the whole host of sea-birds, whose habits are necessarily very different. Many of these sea-birds pass their lives entirely upon the ocean, and sleep at night with

Fig. 172.—FLAMINGOES.

their heads tucked under their wings as they float upon the waves. They feed upon fishes and other small animals, which they snatch while skimming over the water. They go on shore only to raise their young ones, and for this purpose they often select lonely islands and steep, bald cliffs. Here thousands of them raise their young upon the bare rocks, and mingle their screams with the roar of the waves below.

9. In a certain sense, perhaps, these birds are destitute of the charms which we usually expect to find in birds. They seldom take any pains with their nests, and their

Fig. 173.—A FEW SEA-BIRDS.

note is only a shrill cry; but these defects are easily over-looked after once seeing them upon the lonely ocean and learning something of the wild, free lives they lead.

10. **Gulls.**—First of all, there are the gulls, so abundant all over the world. With their strong wings they fly rap-

idly and gracefully over the sea, and when their keen eyes discover a tempting morsel in the water beneath, they make a sudden dive to procure it. These birds meet in large numbers to breed on the rocks, or on the sand-bars at the mouths of rivers and bays. Their shrill cry seems to be half a laugh and half a scream,' and

Fig. 174.—SEA-GULL.

sounds very weird and melancholy when it is heard at night or during a raging storm. The graceful, elegant gulls generally keep quite near the shore, and are not found very far out at sea.

11. **The Stormy Petrel.**—The stormy petrel, on the other hand, is met at great distances from the land. Although the smallest of web-footed birds, and not larger than a swallow, it is very

Fig. 173.—STORMY PETREL.

brave, and flies over the roughest sea with perfect confidence, rising with the billows and sinking with the fall-

12

Fig. 176.—THE ALBATROSS.

ing waves as if delighting in the storm. Watching the petrels is one of the delights of a sea-voyage. To all appearances, the same little flock hovers over the wake of the vessel from day to day, and looks as if it might intend to cross the ocean in company with this particular vessel. These tiny little black-and-white birds are usually called "Mother Carey's Chickens," and they live constantly on the water. Once in a while they make a dive under the waves or float for a moment upon the surface, and then resume their onward flight with as much spirit as before. They never seem to grow weary, neither do they seem to rest.

12. **The Albatross.**—The largest of the swimmers is the albatross, a powerful bird with white body and black wings. It also seems to delight in a fierce gale, and skims over the surface of the ocean without touching it. It is, nevertheless, an expert swimmer when it deigns to light upon the water. An albatross has been known to follow the course of a ship in mid-ocean for many weeks.

13. **The Eider-duck.**—The eider-duck, too, is a real seabird, but it does not fly well, and its habits are peculiar in many ways. Eider-ducks spend the winter in large flocks on the Arctic seas, but when spring comes they select their mates, and the happy pairs swim off by themselves to the shore. The female makes a large, loose nest of dry grass and straw, and lines it with a thick layer of down which she has plucked from her own breast. In

Fig. 177.—THE EIDER-DUCK.

this warm nest she lays from six to ten pale-green eggs, and a supply of loose down is generally placed near by to cover the eggs with when the hen is off in search of food.

14. **The way Eider-down is Obtained.**—Eider-down, as you may know, is very valuable on account of its lightness and softness. The eggs are also valuable, and bring a good price when offered for sale; so it often happens that the natives of these cold countries are on the watch for the nests about this time, and carry them off as soon as they are filled with eggs. This is the way our eider-down is obtained. The mother-bird, in her distress, makes another nest; but her down is gone, and she has nothing to line it with, so her mate is now obliged to strip the down from his breast also. The natives do not disturb this second nest which the male has lined, for if they should destroy the nests too often there might be no eider-ducks to visit their shores by-and-by. Soon after the young ducks are hatched they are taken by their elders to the sea, where they are instructed in the arts of swimming and diving, the mother going down with a little one under each wing.

Fig. 178.—PENGUIN.

15. **The Penguins.**—Very different birds from these are the penguins. Their short legs are placed right at the end of the body, so they can stand only in an erect position, and when they attempt to walk their bodies turn half-way round at each step. Then those paddle-wings, hanging down

at their sides and covered with short scale-like feathers, are entirely useless for flying, but they answer very well in swimming and in scrambling upon the rocks.

16. Penguins live only in the Southern Hemisphere, while their distant cousins, the auks and guillemots, have their home in the North. They spend most of the time in the water, and are often found at a great distance from land. When they come on shore, and stand upright in long lines, exposing their glistening white breasts to the sunlight, they are said to look like an array of soldiers.

Fig. 179.—PELICANS.

17. **The Pelicans.**—We must not forget the pelicans, too, those awkward, ungainly birds that look almost too heavy to fly. But they are not very heavy, after all, for their bones contain a great many air-sacs, and their large heads are nearly all bill. The curious bag that hangs underneath is only a tough, flabby skin, which makes a convenient pouch for the pelican to scoop up fish with, and

carry them off to the shore to be eaten at leisure.   Pelicans are numerous on our Florida coasts, but they are so shy that you will not find it easy to get a good look at them.   On reaching one of the narrow strips of sandy islands that skirt the shore, you will scarcely have caught a glimpse of the pelicans standing on the beach before there is a whirring of broad wings and they are gone.   You may not see them in the air, you may not see them on the water, but after a while you will find that somehow or other they have reached another sand-bar almost out of sight; and this is the way they will serve you again and again.

XLI.

WADING - BIRDS (*GRALLATORES*).

SUB-KINGDOM, VERTEBRATA: CLASS, AVES.

1. **The Heron as a Type of Wading-birds.**—From the long neck and the long, naked legs of the heron we may form a pretty good idea of what to expect of all wading-birds,

Fig. 180.—HAUNT OF THE HERON.

no matter how much they may differ in size. The long, straight toes spread out very far apart, and thus prevent the birds from sinking in the soft mud as they wade about in shallow water near the banks of rivers and marshes. Places like these are the favorite haunts of the herons, and here they stand patiently watching for fishes, frogs, and small reptiles. Their long necks are admirably suited for reaching out to catch such creatures, and their slender beaks quickly seize the prey, seldom missing their aim.

2. Herons are found in all parts of the world, and they form one of the greatest ornaments of our Southern marshes and streams. Their flight, however, is not very graceful. These birds have no tails worth speaking of, so, when flying, they always stretch out their legs behind them, to act as a rudder, while most other birds tuck their legs snugly away out of sight.

3. **Cranes.**—Another large and elegant bird is the crane, which is sometimes over four feet in height, but otherwise not especially remarkable, unless it be for its long migrations twice every year, and for the perfect discipline which is observed on these journeys.

Fig. 181.—CRANE.

Fig. 182.—MARABOU STORK AND YOUNG.

4. **Storks.**—Those of you who have read the interesting stories about the storks that live in European and Asiatic cities, and perhaps have yourselves seen them there, may be surprised to learn that they also are waders. These

12*

city birds seem to have given up their aquatic habits since they came to live in towns, and now they stalk about the

Fig. 183.—STORK'S NEST.

streets amid throngs of people, and are not the least disturbed by them. The presence of the storks in these cities is not only tolerated, but, more than this, the birds are highly valued, because they feed upon garbage and small vermin, and in this way help to keep the streets clean. On account of these services, especial laws have been made in some countries for their protection.

5. Their nests, placed in tall trees, towers, or chimneys, are coarse affairs, loosely built of sticks. In Holland persons sometimes make false chimneys to their houses on purpose for the storks to build on, and that family is considered fortunate that has a stork's nest upon the roof. These dignified birds are especially numerous in the Eastern Hemisphere. They assemble in large flocks before starting on their migrations, and it is a common belief that at such times they are consulting about their intended journey.

6. **The Wood Ibis.**—The beautiful ibises are found in all warm countries. One species, the wood ibis, has gained

for itself the reputation of being very greedy, and not
without good cause, as you will see. With its strong bill
it kills a great many small animals, which form its favorite food. As these victims of the voracious bird lie floating
on the water round about the scene of their destruction,
the ibis swallows as many as it can well take, and then
stands stupidly on the edge of the stream, waiting until
this meal is digested before it is able to indulge in another.

7. **The Sacred Ibis.**—Then
there is the sacred ibis,
which was worship-
ped by the people
of Egypt in olden
times. Perhaps
they loved this bird
because it devour-
ed the serpents
which

Fig. 184.—THE SACRED IBIS.

annoyed them so much, or else because it returned each year at the time of the overflow of the Nile; and the superstitious Egyptians may have thought they were indebted to the ibis for the fertility of the country which results from this overflow. It is at least certain that they were in the habit of embalming the bird with their mummies, and placing curious stiff pictures of it on their monuments.

8. **Shore-birds.**—Among the smaller waders are some of our pretty little shore-birds, whose quick movements are so interesting to watch. Small flocks of these little birds on the beach may be seen running out eagerly after a retreating wave, snatching up tiny fishes and crabs, and hurrying along to gather as many of these dainties as possible before the next wave comes in. Then they all mount rapidly into the air to escape this coming wave, as though exceedingly anxious not to wet their slender toes. Their feast is interrupted but a few seconds, for they soon alight and go through the same performance.

XLII.

RUNNING-BIRDS (*CURSORES*).

SUB-KINGDOM, VERTEBRATA : CLASS, AVES.

1. Ostriches.—First among running-birds we may place the ostriches, whose wings are remarkably small, and quite useless as far as flying is concerned. In fact, these great heavy birds never do fly. Their bones are filled with marrow instead of containing those curious air-sacs which we have noticed in flying-birds; their breastbone has no keel for the attachment of wing muscles, and in several other ways they are deficient in those contrivances which enable most birds to support their own weight in the air.

2. Ostriches like those seen at menageries and zoological gardens are natives of the deserts of Africa and Arabia, and they are the largest of all birds living at the present time. They often measure eight feet in height. Their long, stout legs are covered with scales, and they have but two toes, one of which is much longer than the other, and is armed with a strong claw. The claw sometimes inflicts severe wounds, for the ostrich fights by kicking, and this it does so violently that it can defend itself against large and fierce animals. It also runs with great speed, and few animals can overtake it.

3. The deep, rumbling voice of the ostrich is so much like the roar of a lion that travellers have often mistaken its harsh tones for that dreaded sound. Lions, however,

roam abroad only at night, and this fact aids persons who are familiar with the habits of these two animals in distinguishing one voice from the other. When hiding from enemies, an ostrich is content with poking its head out of sight; this stupid habit often leads to its capture.

Fig. 185.—THE OSTRICH.

4. **The Plumage of Ostriches.**—Judging merely from the elegant ostrich plumes which are offered for sale, persons who have not seen these birds might naturally suppose that they are covered with beautiful feathers; but, on the contrary, their plumage is very thin and scanty. Their heads and necks are almost bare, and it is only the quill feathers of the tail and short wings that supply the ostrich feathers of commerce. You will find, upon examining these feathers, that the barbules do not hook into each other after the usual manner, but each one is quite free

from its neighbors, thus flying apart loosely, and giving to the plume a peculiar softness and beauty.

5. **The Nest and Eggs.**—As to the habits of ostriches in their native homes, the birds go together in flocks consisting of one male and six or seven females. The latter make their nests by scratching a hole in the sand, and each one lays from ten to fifteen eggs. The hens sit upon their eggs, as most birds do; but in those warm countries of which the ostrich is a native, the heat of the sun is so great that the eggs may be left during the day without injury; and as the birds have to roam for long distances in search of food, the nest often remains uncovered for several hours at a time.

6. Some persons, having found the nests thus apparently deserted, have formed the idea that ostriches take no care of their eggs, but leave them to be hatched altogether by the heat of the sun. So far from this being the case,

Fig. 186.—Hunting the Ostrich.

the old birds are extremely careful of them, and the male will sometimes sit upon the nest himself, if he finds one uncovered. After the brood is hatched, those eggs that have not developed, or that have become addled from any cause, are broken open by the mother, who feeds her young birds upon the contents.

7. The Size of the Egg.—One single ostrich egg weighs about three pounds, and is thought to contain as much food as one dozen ordinary hen's eggs. The natives roast these huge eggs in the shell by sticking one end in hot ashes, and making a hole in the upper end of the shell, through which they stir the contents with a stick. They also use the strong, thick shells, after they have been emptied, for drinking-cups.

8. Ostrich - farming. — In the southern part of Africa many persons are employed in a kind of business called ostrich - farming. It is found very profitable to raise ostriches for the sake of their plumes, which are pulled once in every eight months. These always bring a good price, and so, likewise, do the birds and eggs when they are offered for sale. Were it not for caring in this way for the ostriches, and thus increasing their numbers, it is thought they would soon have been destroyed upon their native deserts. Ostriches have also been successfully raised in California within the last few years.

9. An ostrich farm is merely a large grassy enclosure in which are placed a few ostriches. It is customary to hatch the eggs by means of an "incubator," it having been found that the important point in hatching eggs is to keep them constantly at the same temperature as the body of the bird when she is sitting. For this purpose a large chest is used with sliding drawers, between which are vats filled with hot water. The eggs are snugly wrapped in flannel and placed in the drawers, where they remain six weeks before hatching, and great care is taken to imitate the natural method as closely as possible.

10. The Rhea.—A smaller ostrich than the one we have

been speaking of, called the rhea, is found in South America. It lives in flocks on the pampas, and is about one half the size of the African ostrich.

11. **The Emu.**—Another singular bird is the emu, which is found only in Australia. It is nearly as large as the ostrich, but it is not perched upon such long legs, and it

Fig. 187.—EMU AND WOLVES.

has three toes on each foot. This is the only one of the running-birds that wanders about in pairs, and its large eggs are of a beautiful dark green color.

12. The Cassowary.—Running-birds are not numerous, and, as you may have noticed, each species that we have mentioned so far is confined to a limited extent of country. This is also true of the cassowary, which inhabits New Guinea and the Molucca Islands. The black plumage of this bird resembles the hair of a horse's mane, and its head is very conspicuous, with a horny crest on top, while the naked throat is ornamented with red and blue wattles hanging in front. Its wings are extremely small, and they are armed with five naked quills.

Fig. 188.—THE APTERYX.

13. The Apteryx.—Most curious of all the running-birds is the apteryx, which is found nowhere but in New Zealand. It has no tail, and the stumpy wings are quite hidden by its plumage. It digs a deep hole in the ground in which to deposit its one egg, and it frequently runs to this hole for protection when it is pursued. The apteryx seems to be a shy bird, hiding by day, and roaming about at night to hunt for insects and worms.

14. Why Fossil Birds are Rare.—Fossil remains of birds are by no means abundant, perhaps from the fact that most of the rocks in which we find fossils were formed in water. Now how can this fact make any difference in the number of fossil birds contained in the rocks? Let us imagine the case of a bird, in times long past, having fallen into the water at some place that would be favorable for preserving it in a fossil condition; instead of sinking to the bottom, where it might soon be covered with mud and trash, we know that the lightness of the bird's bones would cause the body to float, and that in this exposed situation it would probably be devoured by some hungry animals, and all traces of the bird would thus be wiped out. This is thought to be one reason that fossil birds are so seldom discovered. If our imaginary bird had been a heavy body it would have sunk. Perhaps it would be covered with sediment, and after a very long period of time it might become embedded in solid rock, and thus be preserved as the remains of so many animals have been.

15. **Some Large Fossils.** — Among those fossils which have been found, however, are some ostrich - like birds, much larger than any now living upon our earth. A gigantic bird twice the size of an ostrich is found to have lived in olden times in New Mexico. Fossils of another bird, called the *moa*, twelve feet in height, have also been found in New Zealand.

XLIII.

SCRATCHING-BIRDS (*RASORES*).

SUB-KINGDOM, VERTEBRATA: CLASS, AVES.

1. **Our Common Fowls typical Scratching-birds.** — Our common fowls, scratching in the garden with their strong blunt claws, and calling their little broods to share the dainties they have found there, may furnish an illustration of this class of birds. They pass most of the time upon the ground, as we know, and seldom fly higher than is required to reach their roosting-place, consequently their wings are weak, and they make a peculiar whirring sound when they attempt to fly.

2. The scratching-birds mostly wander about in flocks, one male accompanying each flock, and taking no part in building the nests or helping to raise the young birds. Their nests are usually upon the ground, and the pretty, downy little chicks are able to run about as soon as they leave the egg.

3. Although these birds scratch in the earth for worms, their food consists chiefly of hard grains and seeds which are not easily digested, and they have, therefore, large crops and strong, muscular gizzards.

4. **Turkeys and Pea-fowls.** — Turkeys, which are still found wild in some parts of North America, are scratching-birds ; so also are the gorgeous pea-fowls from India. The male bird, or peacock, as he is called, is celebrated for

his long train of feathers decorated with eye-like spots of metallic colors. The upper feathers composing the train are shorter than those beneath them, and in consequence of this arrangement the spots near their tips are all dis-

Fig. 189.—PEACOCK.

played, making a magnificent array of feathers, long enough to trail for some distance on the ground; but, in stepping, the peacock lifts it slightly to prevent its dragging. Altogether his movements are such as to give an observer the idea that he never wholly forgets his handsome train. To manage and display such a train is certainly no slight matter. Then see him take some stately

position and slowly bend his head from side to side, as if to give the full benefit of the sunlight to his glistening neck and breast, and you will not wonder he has been called a proud bird.

5. When a sudden fancy takes him, the peacock can lift up his tail into the air, and spread out the feathers into a broad fan, and, if he wishes to, he can rattle the shafts of the feathers together so as to make a peculiar noise. The hen looks very meek beside her mate. She has no train at all, and her plumage is a modest brown, while the prevailing colors of the male are blue and green.

Fig. 190.—Argus Pheasant.

6. **Pheasants and their Protective Mimicry.**—Pheasants are also found in Asia, and there they are brilliant birds, although our pheasants are of a sombre color. The golden and the silver pheasants and the argus are all exceedingly handsome. These gayly colored scratching-

birds, such as peacocks and pheasants, are mostly jungle-birds of tropical lands, and their brilliant hues blend well with the bright flowers and the pretty fruits and berries upon which they feed. On the other hand, partridges, guinea-fowls, and scratching-birds in general, have a dingy, spotted plumage, resembling the ground upon which they sit or run, and by this means they escape the notice

Fig. 191.—IMPEYAN PHEASANTS. INDIA.

of birds of prey, to whose attacks they are much exposed.

7. Where Partridges build their Nests.—Partridges, of which we have just spoken, pass the winter in our fields

Fig. 192.—A PARTRIDGE.

and meadows. Early in the spring they select their mates, and begin to build their simple nests close by the root of some tuft of grass or corn-stalks. The nest occupies a slight depression in the ground, and is often placed where overhanging grasses naturally conceal it and the fifteen or twenty beautiful eggs which it holds.

8. The Young Brood.—Young partridges are very active, and they no sooner leave the shell than they start off to run, following their mother like a brood of chickens, and nestling under her wings when she is at rest. It is well they can run, for if they were not able thus to take care of themselves, these tiny little birds in their open home on the ground would stand a poor chance for life. The cunning mother does what she can to ward off danger to her little ones by pretending lameness, in order to entice foes away from her nest. When alarmed, she makes a noise to attract attention, then throws herself on the ground in full sight and flutters along, dragging her wings as if severely wounded. After she has led her pursuers far enough from the nest, her object is accomplished, and she then starts up and flies away in as good condition as ever.

9. Both the parents and the young brood remain together until the following spring, and form what is called a "covey." They are ordinarily found running in open fields or along fences sheltered by thickets, where plenty of seeds are hanging upon the weeds and bushes, and fur-

Fig. 193.—PARTRIDGES IN WINTER.

13

nish abundant food even in winter-time. When it storms, the birds hide away in sheltered nooks, or creep in among thick bushes, huddling close together for warmth.

10. **Quails.** — Quails are smaller than partridges, but their habits are similar, and their clear call of "Bob White" is familiar to most persons. Our quails and partridges both differ from the English birds bearing the same name.

Fig. 194.—THE QUAIL AND HER NEST.

11. **Pigeons.**—Pigeons also are placed among scratching-birds, although, in development, they are greatly in advance of others of the class, for they are good fliers, they have slender toes for perching, and they build their nests in trees. Another striking difference may be noticed in the young birds, a newly - born squab being perfectly naked, and as helpless and uninteresting as can well be imagined. It is, nevertheless, an object of tender care with its parents, both of whom secrete in their crops a soft, milky substance with which to feed their little ones. Taking the bills of the young birds within their own mouths, the parents force this partly digested food down the throats of the squabs.

12. **Wild Pigeons.** — The wild or passenger pigeons of North America associate in large flocks. Millions of these birds sometimes occupy one roost in a forest, and they are described as frequently breaking the branches of the trees with their weight. Passenger pigeons fly hundreds of miles to their feeding-grounds and return at night to

Fig. 195.—A Pigeon-loft.

their nests, each bird finding its own without difficulty. In their migrations they fill the air like a cloud, and, although their flight is rapid, the size of the flock is so great that it is a long time in passing any one point.

13. **Tame Pigeons.**—We are, however, most interested in the tame pigeons, and we find the domestic habits of these birds very attractive. Having once selected their mates, they remain true to each other for life, and both birds assist in building the nest and in sitting upon the eggs. There are never more than two eggs at a time in the nest, but several broods are raised during the year.

14. Altogether, these birds make most satisfactory pets, and the fancy for keeping pigeons dates back to very ancient times. It is estimated that there are at least one hundred different breeds of tame pigeons, and the peculiarities and fine points of these varieties have been carefully cultivated by "pigeon-fanciers." Perhaps you may recognize in the picture of a pigeon-loft on the preceding page some of the familiar breeds, among which are the fantails and the funny pouters, with their crops puffed out so far as nearly to hide the head. All our domestic breeds are thought to have descended originally from the rock-pigeon, which is still found wild in Europe.

15. **The Carrier-pigeon.**—The carrier-pigeon is one of the most popular fancy pigeons, and sells for a high price. This aristocratic bird was employed in olden times to carry messages for kings and princes, but in these days it is kept merely as an ornament, and is carefully shielded from exposure to the weather. The homing-pigeon, a much smaller and plainer bird, is now sometimes trained as a message bearer. The homing-pigeon seems to have a natural facility for finding its way, owing probably to the cultivation of its ancestors for many generations, yet,

notwithstanding this inherited tendency, each young bird requires careful training before it becomes expert.

16. **The Training of Pigeons.**—This training is begun while the pigeons are very young, by taking them a short distance from their cot and setting them at liberty. Rising into the air and looking about them, the birds see their home and fly to it. Day after day they are taken out in the same direction, each time a little farther from home, and they fly back to their cot as before. After a long period of training they become so familiar with the route that they will return from great distances, but this training must have been in

Fig. 196.— Dodo (in the Foreground) and Apteryx.

one direction, and in clear weather. These pigeons do not start home in a fog, and if overtaken by night they stop on the way and do not continue their journey until morning.

17. **The Dodo.**—The curious dodo, which formerly lived on the island of Mauritius, was closely related to the

pigeon. This bird was rather larger than a swan, with soft plumes on its wings and tail, and it was wholly unable to fly. There are no dodos living at the present time, but they appear to have lived until quite recently, and there are two or three old oil-paintings representing these interesting birds. These paintings and a few bones and feathers are all that remain to tell of the former existence of the dodo.

Fig. 197.—BIRDS OF A FEATHER.

XLIV.

CLIMBING-BIRDS (*Scansores*).

SUB-KINGDOM, VERTEBRATA. CLASS, AVES.

1. Climbing-birds.—Next in order after the scratchers is placed a group of birds whose tastes prompt them to leave the ground and climb up into the trees, as parrots and wood-peckers do. The climbing-birds feed upon insects or fruit, and build their nests in holes in the tree-trunks, and, being naturally poor fliers, they prefer to pass most of their time among the branches. Their toes are well

Fig. 198.—PARROTS.

suited to this kind of life, since they are arranged in pairs—one pair on the front of the foot, and the other pair behind (Fig. 199), thus enabling them to grasp the boughs firmly.

2. Parrots. — But parrots do not depend upon their toes alone in climbing; their strong, hooked bills are also brought into service, and they use them in such a way as almost to supply the place of a third foot. The soft, fleshy tongue of a parrot is unlike that of other birds, for it may be moved in any direction, and it is partly on this account that parrots can be taught to imitate the human voice. Their tones, however, are shrill and harsh, and they have generally but few words in which to utter their set phrases.

3. The forests of South America and Australia are especially rich in parrots, and the plumage of these tropical birds is remarkably brilliant. Green is often the prevailng color, but in some species the red tints predominate.

Fig. 199.—FOOT OF PARROT.

4. **The Woodpecker's Search for Food.** — Quite different from the parrots are the active little woodpeckers, which we mostly see standing upright on the tree-trunks, with the stiff points of their tail-feathers pressed against the trunk. Supported in this way, the birds hop up the trees by a succession of quick jumps, making the while a pecul-

Fig. 200.—TONGUE OF WOODPECKER.

iar tapping sound by striking the beak upon the tree. They are now hunting for insects and grubs hidden beneath the bark. When a hollow sound proceeds from this tapping, the bird is encouraged to drill a hole into the bark with its long, straight beak, hoping to find its favorite food. If the search is successful, the woodpecker

Fig. 201.—WOODPECKERS AT HOME.

then puts out its sticky tongue, which is armed near the end with sharp barbs, pointed backward like a fish-hook (Fig. 200), and draws the insect from its lodging-place. This curious tongue is fastened to cartilages which ex-

13*

tend up back of the skull and over to the forehead, and in consequence of this arrangement the tongue can be thrust out some distance beyond the beak.

5. The limbs of apple-trees and maples are sometimes found pierced by rows of little holes, extending in rings,

Fig. 202.—WOODPECKER'S NEST.

one above another, quite around the stem, and it is supposed that the woodpeckers have bored these holes to get at the sap in winter—a plan somewhat similar to that practised by farmers in obtaining the sap from which to make maple-sugar.

6. **The Snug Home of the Woodpecker.**—Many of our

woodpeckers excavate holes in trees in which to pass the winter, taking the precaution to select a spot which is sheltered from rain and snow by an overhanging branch. Such cavities make snug, warm homes, but when spring comes the woodpecker leaves its winter-quarters, and hollows out a new nest in the solid live wood of the tree-trunk, carefully chipping away the inner surface to give it a smooth finish. In Fig. 202 you see the entrances to two nests. That part of the trunk nearest to us has been cut away so as to show one of these nests inside of the tree. The circular entrance to the nest is merely large enough to admit the bird, and after extending some distance within the tree the tunnel turns downward and enlarges into the shape of a long pear. The pure

Fig. 203.—Toucan.

white eggs are laid on the chips at the bottom of this nest.

7. **The Cuckoo an Intruder.**—The European cuckoo has found an easier plan than this, for she builds no nest at all. In fact, she lays her eggs in the nests of other birds, and has the assurance to leave them there to be hatched. Generally she deposits but one egg in a nest, and the young cuckoo is brought up by its foster-parents at the

Fig. 204.—TROGON ELEGANS.

expense of their own young, for this intruder soon manifests the selfish disposition of its race, and slyly tosses the rightful occupants out of the nest. This peculiar habit of appropriating the nests of other birds is possessed also by the cowbunting of America, but our cuckoo makes a nest and raises her young ones in the usual way.

8. **The Toucan.**—The enormous bill of the toucan is not as heavy as one might suppose, for instead of being solid it is hollowed out to contain a great many air-cells. Its shape, too, is admirable for robbing nests and deep holes of the eggs and young birds on which the toucan feeds, tossing them first in the air and catching them as they

fall. Neither is the bill too large to be snugly tucked away among the feathers when the bird is preparing for sleep. Toucans are confined to the tropical regions of South America, where they assemble in large flocks.

9. **Trogons.**—Trogons are also tropical birds, chiefly remarkable for the beauty of their plumage, which is loose and richly colored. They live in the deep recesses of the forests, and sit motionless on the branches watching for insects.

XLV.

PERCHING-BIRDS (*INSESSORES*).

SUB-KINGDOM, VERTEBRATA: CLASS, AVES.

1. **Perching-birds.**—Perching-birds also live in the trees, and build among the branches, displaying great skill in the construction of their nests. They are unlike the climbers, however, in having slender, flexible toes with long claws, well suited to the delicate labor of nest building, and their legs are so weak that these graceful creatures have a dainty way of hopping instead of walking.

2. **Robins.**—We shall find among the perchers some of the most attractive birds of our fields and gardens. The robins, for instance, are always welcome, partly because they come so early in the spring and stay with us until late in the fall, and partly because they prefer to build their nests in the trees and orchards near our homes. Coarse and rough as these nests are, they contain four beautiful greenish-blue eggs, and two or three broods are generally raised in one summer.

3. Robins devour a great many insects, and in this way render valuable service to the farmer. A young robin in the nest is said to consume more than its own weight of food each day, and as this food consists largely of insects, we may form some idea of the great number that must be destroyed in feeding the whole family.

4. **The Baltimore Oriole.**—The Baltimore oriole is a gay bird, with rich orange and black colors. The female bird

Fig. 205.—ORIOLE FEEDING ITS MATE.

has the same markings as the male, although in duller tints, and a pair of these birds forms a handsome ornament to a lawn. The males arrive first from the South, and are joined a week later by the females; at this time they are full of song, and ready to devote themselves immediately to wooing and trying to secure a mate.

5. The Nest of the Oriole.—The hanging nest of the Baltimore oriole, as well as that of the robin, is mostly built near the house. It is suspended from two or more twigs by strings and threads, and through these threads a coarse fabric is woven into the form of a pouch, inside of which is placed the true nest of fine grass or hair.

6. Song-sparrows.—Among the commonest of our summer visitors are the little song-sparrows, whose cheery, melodious note is repeated over and over through the long spring days, from early in the morning until nearly dark. Their snug little nests are generally hidden away in a grassy bank, or placed on some low vine or bush, the male carrying the materials for building it, while the female weaves the nest. He is attentive to his mate, and when their home is completed, and the female sits contentedly upon her eggs, he brings her food and lingers near by to cheer her with his song.

7. The Snow-bird.—The snow-bird is another of the sparrows. These brave little birds stay with us through the cold winter, when most other birds have left, but upon the approach of spring they fly off to Canada or some other northern country to raise their young ones.

8. Yellow-birds.—Our pretty yellow-birds, or goldfinches, also stay with us through the cold weather, but their appearance at this season is so altered by their plain winter garb that they are scarcely recognized as the same bird, and we might easily fancy that the yellow-birds had all left us. When spring comes they again assume their gay coats of yellow and black, enlivening the landscape with their bright colors and delighting us with their sweet songs. These birds are fond of thistle-seeds,

Fig. 206.—FROLIC IN THE SNOW.

and, perched upon the branches of the prickly thistle, they soon tear to pieces the downy balls to obtain the seeds crowded together at the centre.

9. The Crossbill.—That peculiar, crooked beak of the crossbill (Fig. 208) looks like a deformity, but an acquaintance with the habits of the bird shows that its bill is well suited to tear apart tough pine-cones in order to reach the seeds which form its food.   Clinging to a twig of a pine-

Fig. 207.—SUMMER YELLOW-BIRDS.

tree with one foot, it grasps a cone with the other ; then inserting its bill underneath the scales, it pries them apart by a sidewise motion, and scoops out the seeds with its tongue.

10. Crossbills are bright, happy birds. They fly in small flocks, often visiting our gardens and flitting among the evergreens; but their movements are very quick, and they suddenly dart off as unexpectedly as they came.

11. **The Hornbills.**—The hornbills of Africa and Southern Asia are conspicuous for the great size of their bills, which, however, are so filled with air-cavities as to be very light.

12. The nest-building habits of the two-horned hornbill are exceedingly odd, as you will infer from the following

Fig. 208.—THE CROSSBILL.

picture (Fig. 209). Having selected a hollow tree, the female takes her place within the hole and makes a nest of her own feathers, while the male from the outside plasters up the hole with mud, leaving only a small opening for the beak of the imprisoned female. Through this hole she is fed by her mate until her young family is fully fledged, and during this time she requires constant care from her attentive companion to satisfy her ravenous appetite.

13. **Birds-of-paradise.**—Birds-of-paradise live only in New Guinea and the neighboring islands, and here twenty

Fig. 209.—Two-horned Hornbill feeding its Mate.

* different species of these beautiful birds are found. The ordinary birds-of-paradise most familiar to us are admired for the plumes of downy golden feathers growing beneath their wings, and large numbers are killed to supply the

milliner's trade. The natives who capture them usually cut off their legs, and this may have given rise to the

Fig. 210.—Bird-of-paradise.

mistaken notion of olden times that these birds have no legs, that they suspend themselves by their long feathers, and that they never touch the earth while alive.

14. **The Bower-bird.**—Another interesting bird of the Eastern Hemisphere is the bower-bird of Australia. Its chief peculiarity consists in the curious bowers which it builds of closely interwoven branches and twigs, drawn together so as to meet at the top. The entrance is brushed perfectly clean, and decorated with bright pebbles, shells, feathers, little bleached skeletons,

Fig. 211.—PLAY-HOUSE OF BOWER-BIRD.

and other ornamental articles, some of which must evidently have been carried for a long distance.   These
bowers are entirely separate from their nests, and are
used only as play-grounds, where a festive throng assem-
bles, apparently to exhibit their charms to the birds
whose affections they hope to win.   The males strut up

and down in a stately fashion, and do their best to display their fine forms and graceful movements to the females that are quietly looking on.

15. **The Shrike.**—The shrike, or butcher-bird, as it is often called, has a singular habit of hanging small birds, mice, and insects upon the thorns and twigs of trees, as a means of preserving them for future use, having captured more prey than it can possibly eat at one time.

16. **Wrens.**— Then there are the wrens—busy, fussy little creatures, hopping about our bird-boxes with tails erect, and fighting and scolding other birds that are thought to be trespassing upon their possessions.

Fig. 212.—HOUSE-WRENS.

Almost any hole will answer for a nest, and after it is stuffed with twigs and rubbish, six or seven brick-colored eggs are laid in the centre of the mass.

17. **Humming-birds.**—Our pretty little humming-birds belong exclusively to America, and are greatly admired

for their small size, as well as for the metallic lustre of their plumage. Their throats are especially brilliant, and are often adorned with a variety of beautiful colors. The bills of these birds are always long and slender, but it has been observed that they are either straight or curved, according to the shape of the flowers they frequent.

18. No doubt humming - birds are associated in your minds with the flowers about which they flutter. They visit these flowers not so much to obtain honey as to capture the insects which have been attracted by the sweet juices found there. The tongue, however, contains two hollow tubes, and it is divided at the end, thus serving the double purpose of catching insects and of sucking the honey from flowers. This remarkable tongue, like that of the woodpecker, is attached to a cartilage extending over the skull, so that it can be thrust out beyond the beak.

19. The humming sound which you have noticed in the flight of these little birds, and from which they take their name, is produced by the exceedingly rapid motion of their wings. They do not often alight when taking their food, but by beating the air they are able to hover before the flowers long enough to secure their feast; then they dart so suddenly from one blossom to another that we seldom do more than catch a glimpse of their beautiful sparkling colors.

20. Humming-birds live in pairs, and the male defends his little family with much spirit. Indeed, the female herself is not wanting in courage, and she is slow to abandon her nest even after it has been desecrated. These birds sometimes return to the same tree for several seasons in succession, and the young birds appear to stay with their parents until the fall, when they all go South together.

21. The Nest of the Ruby-throat.—The cup-shaped nest of our common humming-bird (the ruby-throat) is a tiny one, made of soft down taken from the stems of ferns. It is then covered with mosses and lichens, so closely resembling the branch on which it is placed that there is not much danger of discovery. Even the keenest eyes might

Fig. 213.—Broad-tailed Humming-birds.

14

mistake it for an old knot or for some roughness in the bark; and but few persons have had the pleasure of finding a humming-bird's nest, with its two white eggs scarcely larger than a pea. When these nests have been found, the attention of the discoverer has in most cases first been attracted by the suspicious behavior of the birds, they having betrayed their own secret by returning so frequently to the same spot.

22. **Some of our Finest Songsters.**—Many of the perching-birds are gifted with fine voices as well as with beautiful plumage, and there are some celebrated songsters belonging to this large group. The European skylark is a great favorite, because of its beautiful song and of the peculiar manner in which it is given. Rising almost perpendicularly from the ground, the skylark sings as it soars, mounting higher and higher into the air until it is lost to sight, although its clear tones may still be heard. The nightingale is another European bird famous for its melodious song. It usually begins its long, quivering strains in the evening and continues to sing through the night. Some of the thrushes, too, are very musical. The mocking-bird is one of these. It is found only in America, and is remarkable for its power of imitating the notes of other birds. Its song at night, however, is natural, and it then pours out a flood of enchanting music.

23. **Bobolinks in the different Characters they Assume.**—Bobolinks arrive in New England early in the spring, and for a few weeks they sing very sweetly. One may see them perched on a twig or spear of grass in the meadow, uttering the while a succession of gay, frolicking notes as they tilt up and down on their slender support. Their nest is probably hidden at the root of some tuft of grass not very far away, but they are so cautious in approach-

Fig. 214.—The Nightingale.

ing it that we shall scarcely discover its position from the
wise little owners.

24. At this season the male is dressed in a mottled coat
of black and white, while the female has one of yellow-
ish brown. Later in summer the male assumes a quiet
garb like that of his mate, and they start off towards the
South in search of good things to eat. They find attractive feeding-grounds among the reeds and marshes of the
Delaware River, and soon grow fat upon the seeds which
abound in such places. They are now called *reed-birds* in

Pennsylvania and New Jersey, and, being considered a great delicacy, they are shot in large numbers to supply the tables of the luxurious.

25. Those birds which escape the gun of the sportsman next visit the rice swamps of Carolina, where they feast greedily upon grains of rice, and pass by the name of *rice-birds*.

26. The character of the bobolink undergoes a complete change during this time. In spring the bird is very musical ; it seems to know that it has arrived among friends, and it becomes tame and familiar. But after starting on its southern journey it loses its refined and musical tastes, grows silent and shy, and gives itself up to the pleasures of appetite.

27. Thus the bobolink in its extended migrations, which are supposed to reach from Labrador to Patagonia, plays the part of three birds, differing in character as well as in appearance.

28. **Swallows.**— Swallows are excellent fliers, as their long, pointed wings and forked tails are both favorable to speed. It is estimated that these birds fly from sixty to ninety miles an hour. They delight in places where insects abound, and here they may be seen during twilight flying in large circles. The sticky glue-like saliva of the swallows serves them a good purpose in nest-building, and they all make use of it to strengthen their nests or to fasten them securely in their places.

29. **Edible Birds'-nests.**—The edible birds'-nests, so popular among the Chinese, are built by a species of swallow. These nests are made of a certain kind of sea-weed, which, when boiled, yields a good quality of glue. The birds first swallow the sea-weed, then deposit from their mouths the material, which has been moistened with their own sticky

saliva, in layers around the edge of the nest, and the whole structure hardens on being exposed to the air. So the birds'-nests are in reality a fine gelatine. These nests are glued upon rocky cliffs and inside of deep caverns on the

Fig. 215.—NEST OF EDIBLE SWALLOW.

sea-shore, and from these places it is extremely difficult and dangerous to gather them.

30. **Bank-swallows.** — Not the least interesting of the swallows are the bank-swallows, which, when seen at all, are generally congregated in large flocks. They dig holes for their nests in perpendicular bluffs of fine sand near

Fig. 216.—HOME OF THE BANK-SWALLOW.

some sheet of water. These holes extend into the bank
for two or three feet, and at the farthest extremity is
placed a loose nest of hay, jauntily lined with a few soft
goose feathers standing upright. One hole is just like another, and they are placed so close together in the bank
that we might suppose it would puzzle the birds to find
their nests, but each one knows its own.

31. **Whippoorwills.**—Whippoorwills are rather dismal

birds, inhabiting secluded spots in the deep woods, and keeping out of sight until night comes on, when they fly forth in search of insects. It is at this time only that we may hear their mournful cry of "whip-poor-will," which is sounded so dis-. tinctly as to be quite startling. They do not build a nest, but merely scratch a hole in the earth near a rock or fallen tree to contain their eggs.

32. **Provision for catching Insects.** — Swallows and whip- poorwills feed upon insects, and no bet- ter trap could be de- vised for catching the tiny prey than is formed by their large mouths, which are moistened by

Fig. 217.—WHIPPOORWILLS.

sticky saliva, and are also furnished with bristles hanging from the roof of the mouth. As these birds fly about with their mouths wide open, the insects become hope- lessly entangled among the sticky bristles, and are thus prevented from escaping.

XLVI.

BIRDS OF PREY (*RAPTORES*).

SUB-KINGDOM, VERTEBRATA: CLASS, AVES.

1. **Hawks.**—As hawks are among the commonest of our birds of prey, let us take them as our example, and notice how admirably the strong, curved beak and the sharp

Fig. 218.—HAWK AND HUMMING-BIRDS.

claws (as shown in Fig. 219) are suited for seizing their victims and tearing the flesh from their bones.   These birds have also large, strong wings, and their flight is rapid and powerful.

2. Hawks usually fly quite low, and we may recognize them at a distance by their habit of flapping their wings rapidly for a while, and then soaring without apparent effort for an equal length of time.

Fig. 219.—CLAW AND BEAK OF BIRD OF PREY.

3. **The Flight of the Eagle.**— The movements of the eagle are particularly easy and graceful. Its strong wings bear it onward and upward to a great height in the air, and it forms one large circle after another, as if delighted with its own performances.

Fig. 220.—THE GOLDEN EAGLE.

4. **The Golden Eagle.**—The golden eagle may be seen circling in this majestic manner about the lofty peaks of mountain regions, where it places its nest, or "eyrie," on

14*

Fig. 221.—Eagle's Nest.

the highest and most inaccessible cliffs. The nest is a very rude one, built of large sticks and branches roughly heaped together; still it becomes the family residence, and is used year after year, seldom being abandoned for a new nest. Neither do the parent birds desert each other when their young ones are grown; but having once chosen their mates, they continue true to them the rest of their lives. With eagles as with other birds of prey, the female is larger than the male.

5. **A large Supply of Food required.**—For food the golden eagle seizes large birds, rabbits, fawns, and sometimes sheep and lambs, and carries them off bodily to its secluded nest, where they are torn in pieces and devoured. In this way a pair of eagles and two or three young eaglets consume a large amount of food; and as the bird sails through the air, its keen eye is directed towards the earth in search of fresh supplies. Small animals, either living or dead, are quickly discovered, and the great bird pounces upon them with its strong claws.

6. Shocking as this destruction of life may at first appear, we must not lose sight of the fact that in seeking such food as birds and beasts of prey do, they are only following the instincts of their nature, which requires animal food, and they therefore cannot be regarded as cruel in their habits.

7. **The Bald Eagle.** — The white-headed eagle is often called the bald eagle, because, at a distance, the white feathers on its head have somewhat the effect of baldness. This eagle is fond of fish, and in order to gratify this fondness it frequents the sea-coast, and builds its nest in the forks of a large dead tree. It occupies the same nest from one year to another, making fresh additions to it each season, until it grows to be a huge structure.

8. **Encounters with the Fish-hawk.**—The top of a tall tree upon the coast affords a good view of both the sea and shore, and here the eagle sits and watches for prey. Rather than do its own fishing, the white-headed eagle prefers to attack the fish-hawk, and rob it of fish already caught from the ocean. In the encounters between these two birds, the fish-hawk often drops the fish it is carrying

Fig. 222.—The Fish-hawk and its Nest.

to the shore, and the eagle darts after the falling booty
with such speed as to catch it before it reaches the water.

9. **The Vulture.** — The neck of the vulture is mostly bare, but at the lower part there is a loose fold of skin which is covered with soft feathers, and under this warm fold the neck and greater part of the head can be drawn as a protection from the cold. Vultures are cowardly birds, feeding greedily upon carrion until they become almost stupefied. They seldom attack living animals, but they have been seen to sit and watch those that were sick or enfeebled until life was gone before beginning their feast. Even these unattractive birds have their use in nature, for they remove decaying animal matter, the odor of which they can discover at a great distance. In tropical climates their services are especially valuable, as the dead animals thus removed would otherwise become injurious to health.

10. **The Condor.** — The throat of the condor, instead of being bare like that of the vulture, is ornamented with a ruffle of showy white feathers. This huge bird is found only among the Andes Mountains, and here it soars in uninterrupted circles above the high peaks. Its movements are very imposing, and it is said to soar for half an hour at a time without once flapping its wings.

11. **The Owl.** — What a wise-looking bird the owl is, with its large round eyes gazing directly in front of it! The circle of feathers which surrounds the eyes adds still further to the gravity of its appearance, while its loose plumage, extending all the way down to the tips of the toes, is extremely soft and pretty. There is a great advantage, too, in these loose, fluffy feathers, for they render the owl's flight almost noiseless, and consequently these dignified hunters are able to approach the cautious mice and timid little birds in the darkness without being heard. The small animals upon which owls feed are swallowed

whole, and the indigestible portions, such as the feathers, bones, etc., are afterwards thrown out from the mouth in the form of small round balls.

12. **The Nocturnal Habits of the Owl.**—Most owls hunt their prey only at night, or during twilight, at which time

Fig. 223.—The Owl.

they seem to have no difficulty in seeing their victims. Their nests are generally placed in the hollow of an old tree, and in this snug retreat they spend the daytime, venturing out only under cover of the darkness, and occasionally breaking the stillness of the night with their doleful notes.

XLVII.

THE ORNITHORHYNCHUS.

SUB-KINGDOM, VERTEBRATA: CLASS, MAMMALIA.

1. **The Ornithorhynchus.**—Let us now leave the subject of birds and visit in fancy the island of Australia, which is the home of many singular animals as well as plants. Here we shall find the streams and pools frequented by

Fig. 224.—ORNITHORHYNCHUS.

a small animal called the ornithorhynchus, or duck-bill, which is found nowhere but in Australia and the neighboring island of Tasmania.

2. **Resemblance to both Birds and Quadrupeds.**—The or-

nithorhynchus bears a curious resemblance to both birds and quadrupeds, and has, on this account, attracted much attention. The body is like that of an otter, covered with short, brown fur, while the head is supplied with a large, flat beak, much the same in shape as a duck's beak. Altogether this animal is so peculiar that the first descriptions of it were scarcely believed to be true, and when a stuffed specimen was taken to England, persons suspected that a joke had been practised upon them, and that the bill of some Australian bird had mischievously been fastened to the head of a quadruped.

3. **The Habits.**—When swimming, the ornithorhynchus shows only its head above the surface of the water, and it obtains its food of worms and insects in the same manner as the duck, by thrusting its bill into the mud. It is a timid creature, preferring twilight to the glare of day, and taking fright very easily if any attempt is made to capture it. It dresses and pecks its fur with great care, and when asleep it rolls itself up so snugly that one might almost mistake it for a ball.

4. The broad tail and short legs are no doubt helpful in swimming, and the web which unites the toes tells its own story so plainly as to need no interpreter.

5. But the ornithorhynchus is a burrowing animal as well as a swimmer, and although this web extends beyond the claws on the forefeet, yet it does not interfere with digging in the earth, because when any burrowing is to be done the web can be folded back out of the way.

6. **The Underground Nest.**—The nest of this curious animal is under ground near a stream of water, and there are two passages by which it may be entered ; the opening of one passage is under water, while the other opening is in the bank above the surface of the stream. This

nest is lined with grass and weeds, and here, at the end
of the burrow, which is sometimes forty feet in length,
the tiny young animals are raised.

7. **The Ornithorhynchus is a Mammal.**—The ornithorhynchus must serve as our first example of the great group of
Mammals simply because it is lowest in the scale, and not
because it is a good illustration.   In fact, it is a poor rep-
resentative of the class, and you will see that it presents
some strange contradictions.   It has a furry coat, to be
sure, as we have already noticed, and this is one great
peculiarity of Mammals—that they are all more or less
covered with hair at some time in their lives.

Fig. 225.—Burrow of Ornithorhynchus.

8. **The Young Ornithorhynchus begins Life differently from most Mammals.**—Another important peculiarity of Mammals is that their young are born alive, and are nour-

Fig. 226.—Ornithorhynchus and Porcupine Ant-eater.

ished for a time by the mother with milk secreted in the mammary glands. We should, therefore, naturally expect the young ornithorhynchus to begin life as other Mammals do; but the truth of the matter is, these odd animals lay eggs after the manner of birds and reptiles, and thus confuse our attempts at classification. Two eggs are laid at a time in the nest, and when first hatched these little creatures with very long names are quite blind and helpless.

9. **Characteristics of Mammals.**—We have just learned that all Mammals have a covering of hair at some period of their existence, and that their young are born alive,

except in the case of the ornithorhynchus and the porcupine ant-eater, both of which lay eggs, and both belong
to Australia. Following the description of Mammals still
further, we may state here some facts with regard to
structure which apply to all members of the group, and
which will not, therefore, need to be repeated hereafter.

Fig. 227.—HEADS AND FEET OF DUCK, ORNITHORHYNCHUS, AND PORCUPINE
ANT-EATER.

The thorax and abdomen of all Mammals are separated
from each other by a muscular partition called the dia-
phragm. The thorax contains the heart and lungs within
its walls, while the abdomen contains the greater part of
the alimentary canal, the liver, the kidneys, and other organs. There is a perfect double circulation, the same as

in birds, and the senses are highly developed. Most animals belonging to this class have an external ear for collecting the vibrations of sound, and the eyes are protected by two lids which are fringed with eyelashes.

10. The usual number of toes (or of fingers, as the case may be) possessed by Mammals is five, but we find the number sometimes varying from one single toe to a full set of five, each one of which is furnished with a nail, a claw, or a hoof.

11. **The Teeth of Mammals.**—Most Mammals are supplied with teeth, which grow from separate sockets and form but a single row in each jaw. These teeth differ from ordinary bone in being denser and containing less animal matter; they are also covered with a hard substance called enamel, which helps to preserve them from decay. Teeth differ so much in size and shape that they have received different names according to their position in the mouth, and they are generally spoken of as incisors, canines, and molars. The sharp-edged front teeth, used in cutting food, are the incisors. The pointed "eye-teeth," which are fitted for seizing and tearing prey, and which are conspicuous in all carnivorous animals, are called canines. The molars are the grinding teeth, and their shape varies in accordance with the habits of the animal.

12. **The Growth of the Hair.** — It is also interesting to know that hair, which is so characteristic of Mammals, grows from small sacs in the skin much in the same way as feathers, except that it does not split up in the process of development. This difference in the manner of growth seems but slight, yet it produces very different results, and the many kinds of fur and wool with which Mammals are clothed bear little resemblance to the plumage of birds.

.XLVIII.

KANGAROOS AND OPOSSUMS.

SUB-KINGDOM, VERTEBRATA : CLASS, MAMMALIA.

1. **Kangaroos found only in Australia.**—Kangaroos likewise belong to Australia, and, as is the case with the ornithorhynchus, they are never found in any other part of the world. Upon this island, however, they are very abundant. Forty species of kangaroos are known to exist here, and these species differ greatly in habit, some being fitted to live on the desert, while some delight in climbing trees in the forest, and others take naturally to the rocks and plains. An equal diversity may be noticed with regard to their size, which varies from the height of a rabbit to that of a large sheep.

2. **The Great Kangaroo.**—From these different species we shall select as our subject the great kangaroo, which is represented in the picture (Fig. 228), and which lives in large herds on grassy plains. The front parts of the body are strangely out of proportion to the back, and the short forelegs and the delicately formed head, with its mild countenance and soft eyes, look quite unsuited to an animal having such stout hind limbs.

3. **The Movements of the Kangaroo.**—The habitual gait of the kangaroo is a succession of long leaps, which bring into play the muscles of the strong hind limbs, and which have, in fact, tended to produce their extraordinary devel-

Fig. 228.—The Home of the Kangaroo.

opment. In this violent exercise the powerful tail has its part to perform also, and by its assistance the kangaroo is able to make enormous leaps in rapid succession. When the animal is feeding upon the grass by the way, it walks on all four legs, and its motion is then slow and ungraceful.

4. The Pouch of the Kangaroo.—But the distinctive feature of the kangaroo, and the one which especially interests naturalists, is a curious pouch which the females have for carrying their young ones.

5. The Helpless Young.— At the time of their birth young kangaroos are extremely weak and helpless, and the mother soon lifts her tiny babies into the pouch which

Fig. 229.—KANGAROOS.

is formed by a fold of the skin of her abdomen. Within this pouch are situated the teats, and although the little creatures are too feeble to obtain their nourishment by sucking, they take hold of one of the teats and remain attached to it night and day, while the mother now and then feeds them by forcing the milk from her own body into their mouths.

6. The tender young animals are carried in this warm cradle until they are strong enough to depend upon themselves. As they grow larger, they occasionally stick their little heads out of the pouch to look around them, and to nibble at the grass within reach. Later on they jump out of the pouch, and scramble back again as they please, always taking refuge there upon the slightest alarm.

7. **Marsupials.** — Animals possessing these remarkable pouches are called marsupials. With the single exception of the opossum of America, marsupials are confined exclusively to Australia and the islands near it. And it is a curious fact that nearly all the Mammals found here belong to this interesting order of marsupials.

8. **The Virginia Opossum.** — The Virginia opossum is about the size of a large cat. The inner toe of the hind foot can be folded against the other toes, somewhat like a thumb, and this arrangement, together with the prehensile tail, makes the opossum an expert climber. It lives among the thick forests, and is sometimes seen hanging from the boughs by its tail, or swinging from one tree to another by catching hold of the neighboring branches.

9. The opossum sleeps through the daytime in its burrow, and starts out at night to play and to search for food, which consists of fruits, small quadrupeds, birds, eggs, etc. It is a cunning robber of poultry-yards, and, being exceedingly wary and cautious in all its movements, it is not easily caught in its depredations. When the opossum is attacked, its first impulse is to escape by climbing into a tree ; but failing in this, it rolls itself up in the shape of a ball and pretends to be dead, acting its part so well as often to deceive even the dogs. The expression "playing 'possum " has no doubt originated from this favorite trick.

10. **The Young Opossums.**—Young opossums are said to weigh only about a grain at the time of their birth, and to show but little indication of the shape to which they grow afterwards. When they are old enough to leave

Fig. 230.—VIRGINIA OPOSSUM.

the pouch, they sometimes curl their slender tails around the strong tail of their mother, and, huddled together upon her back, they cling to her in this odd fashion as she moves about among the branches of the trees.

15

XLIX.

SLOTHS, ARMADILLOS, AND GREAT ANT-EATERS.

SUB-KINGDOM, VERTEBRATA : CLASS, MAMMALIA.

1. South America the Home of the Edentata.—Australia, as we have seen, has its ornithorhynchus and its kangaroos, and New Zealand its wingless birds. Another ex-

Fig. 231.—Sloth.

ample of this partial distribution of animals is found in South America, which is exclusively the home of the sloths, armadillos, and great ant-eaters. All of these sluggish animals belong to the order Edentata, so called from the fact of their having no true teeth.

2. **The Peculiar Habits of the Sloths.** — The strangest thing about the sloths is that they pass their whole life hanging from the branches of trees with their backs downward, as seen in the picture (Fig. 231). The structure of the body is especially fitted for this peculiar posi-

Fig. 232.—ARMADILLO.

tion, and scarcely admits of any other; so they hang there day and night, even when asleep, trusting to the grasp of their strong, curved claws.

3. They feed upon the leaves and young shoots of trees, and rarely descend to the ground if they can avoid doing so. In a dense forest they can readily swing from the branches of one tree to another in order to find a fresh supply of food; and in thus changing their abode they often take advantage of a time when the boughs are swayed to and fro by the wind. But so great is their aversion to coming to the ground, that when the trees are standing

too far apart to be reached in this ingenious manner, the sloths will devour every particle of foliage on the tree upon which they are hanging before they leave it to climb into another.

Fig. 233.—THREE-BANDED ARMADILLO.

4. Their Feet not fitted for Walking.—These singular animals are clothed with dull, thick hair, much the color of the bark and moss; so they are quite inconspicuous among the leafy branches, and are safer in this retreat than on the ground. Here they have great difficulty in walking, as their curved feet and long claws prevent their treading fairly on the bottom of the foot. They are therefore obliged to step on the side of the foot, and the sole is turned towards the body. Owing partly to this defect, and partly to the fact that their fore limbs are much longer than the hind ones, their gait is extremely slow and laborious. Seen under these circumstances, the sloths appear to deserve the name they have received; but when really at home in the tree-tops of their native forests, they climb among the branches with great ease, and their movements are not then particularly slothful.

5. **The Armadillo lives in a Coat of Mail.**—Armadillos, on the other hand, are burrowing animals, and their strong claws are used for digging. But they are chiefly remarkable for their thick coat of mail, which consists of hard, bony plates united at their edges. One of these plates covers the head, another the shoulders, and a third protects the hinder parts of the body, while between these last two shields several movable plates of the same bony material extend like bands around the body, and allow it to bend freely.

Fig. 234.—ARMADILLOS ROLLED FOR PROTECTION.

6. **How Armadillos protect Themselves.**—When these animals are attacked they burrow rapidly into the earth. Some species roll up into a ball, as shown in Fig. 234, thus securely protecting themselves. At such times the head and tail are drawn closely together and tucked snugly into a little crevice, where the two extremities of

the shell meet, and the result is a hard, solid ball, which may be rolled over and trampled upon without injury.

7. **The Great Ant-eater.**—Still another phase of life is shown by the great ant-eater, an animal four or five feet

Fig. 235.—GREAT ANT-EATER.

in length, with a large bushy tail, which is sometimes thrown over its body as a shade from the sun. Its long jaws are covered with skin, except at the end, where there is an opening through which the worm - like tongue is thrown out. The ant-eater as well as the sloth has curved claws, and it also walks upon the side of its foot.

8. This curious animal feeds almost entirely upon white ants. It tears open their nests with its strong claws, and as the inmates rush forth in alarm, it thrusts out its long, sticky tongue into their midst and then swallows the multitude of ants adhering to it. This operation is repeated

again and again with surprising rapidity, and large quantities of ants are thus devoured.

9. Fossil Remains of the same Type found in South America.—Not only is this order of toothless animals peculiar to South America in the present day, but here are found, likewise, most of the fossil remains of extinct animals of this type. Some of these fossils are interesting from their great size. The megatherium, for instance, was an immense sloth-like animal, eighteen feet in length with, bones as massive as those of an elephant, and the glyptodon resembled a large armadillo, except that it had no transverse bands in its shield. The body was covered

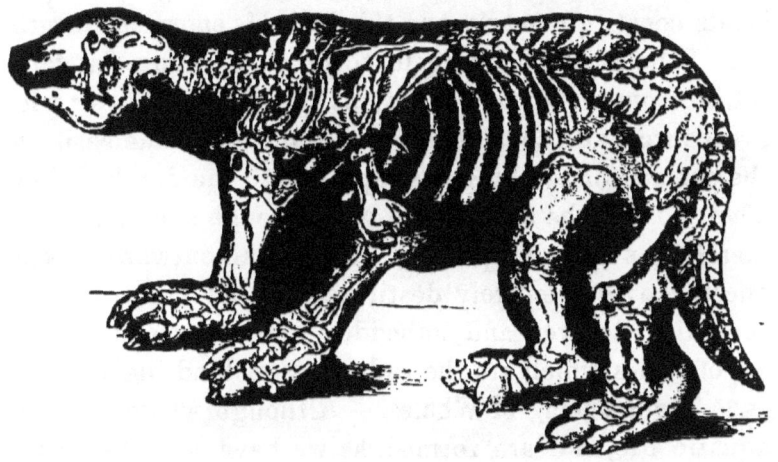

Fig. 236.—MEGATHERIUM.

with one large plate of bone shaped like a turtle's shell, and the glyptodon must, consequently, have been unable to roll itself up as the armadillos of our own time do.

L.

WHALES.

SUB-KINGDOM, VERTEBRATA : CLASS, MAMMALIA.

1. **Whales.** — The largest animals now living are the whales, huge inhabitants of the ocean, which sometimes reach the length of seventy or eighty feet, and whose heads constitute nearly one-third of this enormous length.

2. Whales are fish-like in form, with strong, flat tails, set horizontally in such a way as to strike the water with great force, and thus enable the animal to come easily to the surface, or to plunge as rapidly into the depths below. Their fore limbs are incased with a tough skin, and serve merely as swimming-paddles, while, to all outward appearances, they are entirely destitute of hind limbs. Under the skin, however, and imbedded in the flesh, there is a set of bones which are the rudiments of hind limbs.

3. **The Spouting of Whales.** — Although whales lead an aquatic life, and are formed, as we have seen, for swimming, still, they are true air-breathing Mammals, and they are obliged to come to the surface once in a while to fill their lungs with a fresh supply of air. It is at these times that the curious " spouting " or " blowing " occurs; but the representations of this interesting performance have been greatly exaggerated, and instead of spouting large streams of water, as we have been led to suppose, they merely send up a delicate fountain of spray from the nos-

Fig. 237.—WHALE, WITH ITS YOUNG CALF.

trils, or "blow-holes," which are situated on top of the head, and which are provided with valves to keep out the water.

4. Upon rising to the surface of the ocean, whales begin to drive the air from their lungs before they reach the

15*

top, and the water which is above the head is forced upward by the violent expiration. In addition to the display which is thus produced, the watery vapor from the
lungs is suddenly condensed on coming into the cold at-
mosphere, and these two causes combine to form the pleasing and ever-welcome spectacle of a fountain at sea.
This spouting is accompanied by an explosive sound,
somewhat like that of a large wave breaking upon the
shore, and as it is necessarily repeated at certain intervals,
the whale is unable to conceal its whereabouts even when
closely pursued.

5. **Whales strongly attached to their Young.**—It is believed that whales live to a great age; but as it is not
possible to obtain any facts upon the subject, this point,
as well as many others connected with their life in the
boundless ocean, must remain in doubt. They are said,
however, to select mates, and to be strongly attached to
them and to their young, and whalers tell us that the
mother often swims or floats upon the rolling waves
holding one flipper tenderly over the back of her calf.

6. **The Greenland Whale valuable for its Oil.**—The Greenland whale, which lives in the Arctic seas, is the one sought
by whalers for its oil; hence it has received the name of
"right whale." The blubber, from which the oil is obtained, is a layer of fat connected with the skin, and covering the animal, in some instances, to the depth of two
feet. This thick layer of blubber serves a double purpose, and gives buoyancy to the massive body of the
whales at the same time that it protects them from the
extreme cold of the icy waters.

7. **Whalebone.**—The whalebone of commerce is also
taken from the right whale. This valuable article grows
in broad plates (Fig. 239) which hang from the roof of the

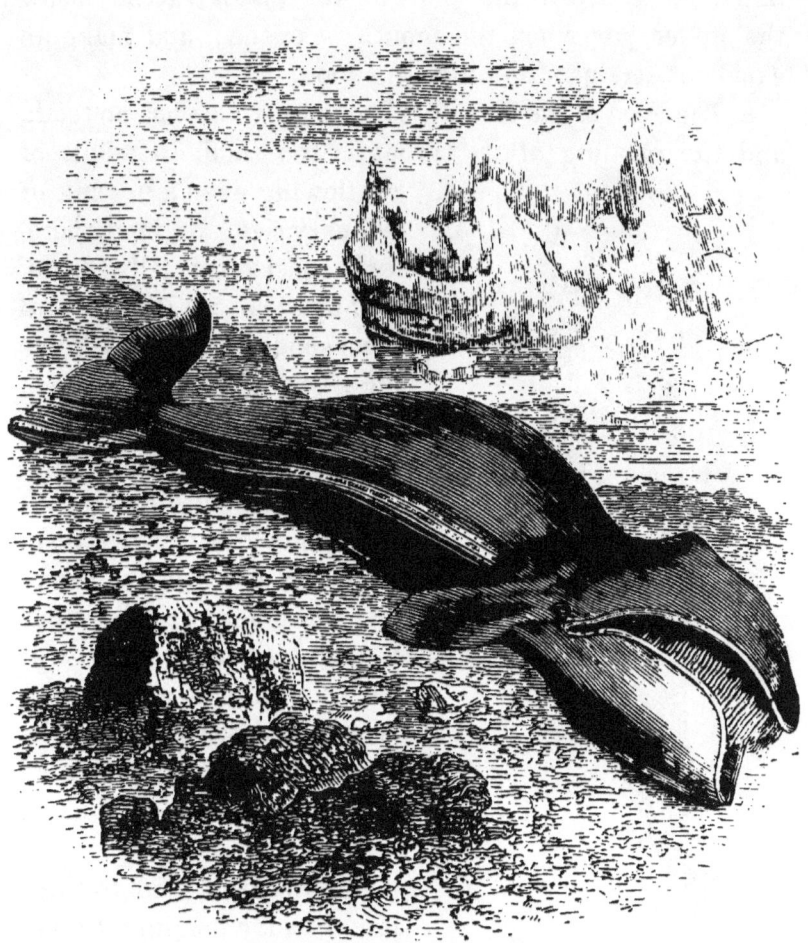

Fig. 238.—GREENLAND WHALE.

mouth, and there are sometimes as many as three hundred of these whalebone plates suspended side by side. The outer edge of the plates is smooth and unbroken, but the inner edge, towards the middle of the mouth, is fringed with frayed-out whalebone fibres, so that the roof of the mouth has the appearance of being covered with coarse hairs.   The brush - like ends of the plates extend below the under jaw when the mouth is opened, and make an excellent strainer for collecting food.

8. **The Food of the Whale.**—The right whale has no teeth, and the opening of its throat is too small to admit of swallowing even a herring of ordinary size. Its food, therefore, consists of jelly - fishes, ctenophora, mollusks, and other small animals which live together in great shoals in the Arctic seas. While feeding, the whale swims through these shoals with its mouth standing wide open; a stream of water constantly passes into the mouth and flows out at the sides, bringing with it the tiny animals and leaving them entrapped in the whalebone fringes. When they have been collected thus in sufficient quantities, they are swallowed from time to time, and you may imagine it takes a large number of such jelly-like creatures to satisfy the appetite of these monsters.

Fig. 239.—WHALEBONE.

9. The right whale is mostly found alone or in pairs, unless it be when larger numbers are attracted to good feeding-grounds.

10. **The Rorqual.**—The rorqual differs from the right whale in having its skin ridged with deep furrows. It is

generally this whale which is caught on our shores, as it ventures farther south than the right whale.

11. **Sperm-whales.**—Sperm-whales frequent the tropical seas, and here they live in great schools. They may be recognized by their large, square heads, which have a single blow-hole near the extremity of the snout. They have none of the curious whalebone plates we have just been studying about, but the narrow lower jaw is furnished with large, conical teeth, slightly curved, and when the mouth is closed, the teeth fit into cavities in the upper jaw.

Fig. 240.—SPERM-WHALE.

12. **Spermaceti.**—These whales are captured for the purpose of obtaining spermaceti, which is a fatty substance in a semi-fluid state, but which, on being taken from the animal, hardens as it cools. The large head is partly occupied by a cavity containing spermaceti, and other cavities throughout the body are also filled with it.

13. **Ambergris.**—Ambergris is another valuable product of the sperm-whale. This waxy substance has an agreeable odor, and it is used in the manufacture of perfumery.

Fig. 241.—DOLPHINS.

It is found in large quantities in the intestines of the whale, but at times floating masses of this peculiar substance are washed on shore, and it is then easily gathered for sale. An interesting point with regard to ambergris is that it is thought to result from slight injuries to the intestines received from the "parrots' beaks," which you will remember as being characteristic of the cuttle-fish family. Cuttle-fishes form the principal food of the sperm-whale, and when these parrots' beaks are swallowed they are supposed to produce in the alimentary canal an irritation which causes the formation of ambergris.

14. **Dolphins and Porpoises.**—The dolphins and porpoises, so common in all seas, are much smaller than the true whales, and their playful capers are highly entertaining. They will follow a vessel in large companies, often leaping out of the water, and frolicking and tumbling about under her very bows. When chasing their prey they sometimes pursue schools of small fishes with such eagerness as to follow them into our bays and for some distance up the rivers. The dolphin is more elegant in form than the porpoise, and may be known by its long snout.

15. **The Narwhal remarkable for its Tusk.**—Another whale requiring our attention is the narwhal, or sea-uni-

Fig. 242.—THE NARWHAL.

corn, which is remarkable for its one great tusk. This strong weapon is possessed only by the males, and it is in reality the left upper incisor, grown to a prodigious length. It projects from the upper jaw, straight forward

in the line of the body, and it keeps on growing throughout the whole life of the animal, until it sometimes measures eight or ten feet—a goodly-sized tooth, which is spirally twisted, and tapers to a point. Its companion, the
right incisor, is only a rudimentary tooth, and does not
often grow into view. The ivory of the narwhal's tusk
is very valuable, as it takes a fine polish, and retains its
beautiful whiteness for a long time.

LI.

HORSES.

SUB-KINGDOM, VERTEBRATA: CLASS, MAMMALIA.

1. Horses known chiefly as Domestic Animals. — Horses have so long been associated with man, and employed so exclusively in his service, that we scarcely realize there

Fig. 243.—HORSES.

was once a time when these noble animals were free and unrestrained. There are some wild horses at the present day, it is true, yet it is believed that all of these have descended from tame horses which escaped from their masters.

2. Horses brought to America by the Spaniards. — The history of the horse on our own continent, so far as it is known, is exceedingly interesting. It shows that at the time of the discovery of America there were no horses here, and that they were afterwards brought into the country by the Spaniards during the Mexican wars.

3. According to the accounts of this conquest, the natives of Mexico were greatly astonished to behold the invaders upon horseback. Not only was the beautiful animal itself wholly unknown to them, but their surprise was further increased by the remarkable sight of a man seated upon its back.

4. Proofs of their Former Existence in America. — Notwithstanding the fact that horses were then unknown in America, still the fossil remains which have been found in the western part of the United States prove that horses existed in the New World in very early times. Therefore, for some good reason which is not understood, they must have died out upon this continent before the arrival of the Europeans upon our shores.

5. Descent from a Small Horse with Four Toes. — These interesting fossils likewise show us that the horses of that far-off time were curious little animals, very different from the graceful, elegant horses of our own day. The earliest of these creatures yet found was a small animal, only about the size of a fox, with four well-developed toes on the fore-foot and rudiments of the fifth toe.

6. Gradual Loss of Superfluous Toes. — Since that time the horse has gradually increased in size and lost its superfluous toes, and naturalists now have the satisfaction of tracing its descent by means of these fossils through the intermediate four-toed and three-toed forms, down to the horse of the present day with its one perfect toe.

7. These changes must, of course, have occurred by easy stages. The side toes, no doubt, gradually diminished in size, and at the same time the middle toe grew larger and

Fig. 244.—GROUP OF HORSES.

stronger, supported by the solid hoof, which is merely a very thick nail.

8. **The Modern Horse a One-toed Animal.**—Horses, therefore, as we know them, walk on this long middle toe, which is covered with the strong hoof, and forms what is

generally spoken of as the foot. What we call the horse's knee is in reality the heel, and under the skin just below the heel may yet be found two slender "splint - bones," which are remnants of the lost toes of the ancestor of our modern horse.

9. **Valuable Service to Man.**—Horses are remarkably intelligent and docile, showing a strong memory for places. They yield themselves wholly to the service of man, often entering with enthusiasm into the work assigned to them; and we can scarcely estimate the assistance they have rendered him in the spread of civilization throughout the world. Each of the various breeds of horses is suited to some special kind of labor; and we may notice that while one breed excels in speed, and another in strength, others are valued for their powers of endurance.

10. **Small Horses in Cold Countries.**—In cold and stormy regions the horses are apt to be small and shaggy, as is the case with the ponies' of the Shetland Islands. These ponies are exposed to bitter cold in their native island home, and they need all the protection which is afforded by their thick, shaggy coats.

11. **Wild Horses go in Troops.**—Horses in a wild condition are in the habit of congregating in large troops, which are led by one male. Fierce conflicts occur between the males to secure this leadership, and the unsuccessful ones are sometimes driven off from the flock into a solitary life. They fight by throwing the fore - feet with great force upon their enemies, or by kicking violently with the hind-feet.

12. **The Zebra.**—Closely related to the horse is the zebra, which is conspicuous for its slender limbs and beautifully striped silky hair. It is altogether one of the most elegant animals, but its disposition is vicious, and it is not

Fig. 245.—Zebras.

easily tamed. Zebras are natives of the southern part of Africa, where they graze in large herds upon the grassy plains.

LII.

DEER.

SUB-KINGDOM, VERTEBRATA: CLASS, MAMMALIA.

1. **The Deer.** — The beautiful deer, with their slender limbs and small heads proudly erect, are general favorites, but to appreciate them fully they should be seen in their natural wild state. The forests of most countries, except Australia, are adorned with their elegant figures, and as they are timid animals they have a tendency to live together in flocks. They run rapidly, and are exceedingly graceful in all their movements.

2. **Antlers unlike the Horns of other Animals.** — Perhaps the most distinguishing feature of the deer family is their large horns, or antlers, as they are called, which are possessed by the males only, and are quite different from ordinary horns. The antlers are solid, and are branching in form. Regularly, at the end of each year, they fall off and are replaced by new ones, whereas the horns of most animals are hollow, and grow around a bony core which is part of the skeleton itself; consequently, these horns are never shed, and one set lasts during a whole lifetime.

3. **The Growth of the Antlers.** — The antlers of the deer are in their finest condition in the autumn and winter, but towards spring the stag is observed to rub his head restlessly against the trees, as if to rid himself of an uncomfortable burden, and when the antlers finally drop off he

is deprived not only of his chief ornament, but also of his means of defense. Tender little knobs push up in place of the cast-off antlers, and he is soon furnished with a fine new pair.

4. These knobs are covered with a velvety skin, which is richly supplied with blood-vessels, bringing material to

Fig. 246.—STAG, OR RED-DEER.

build up the new antlers. The antlers grow rapidly, and send out branches at the proper points, so that within four or five months another pair of horns takes the place of the old ones, and the deer is again fully equipped, and eager to try his strength with his fellows.

5. When the antlers have reached their full size, the

blood - vessels to which we have alluded are gradually closed, and as the velvety skin is thus deprived of its nourishment, it dries and peels off, leaving the strong new

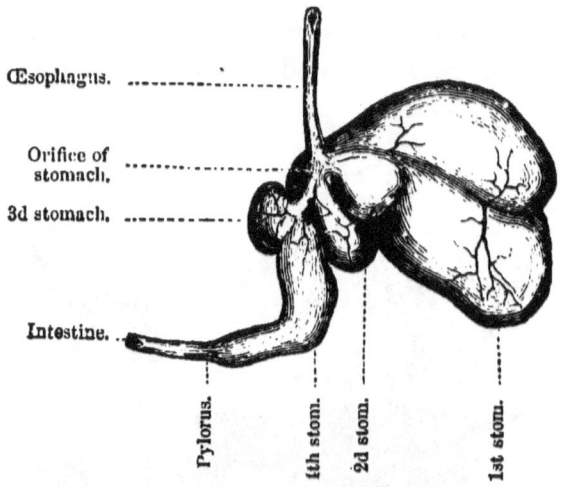

Œsophagus.

Orifice of stomach.

3d stomach.

Intestine.

Pylorus.

4th stom.

2d stom.

1st stom.

Fig. 247.—STOMACH OF A RUMINATING ANIMAL.

horn uncovered.   The antlers usually gain one additional prong with every new growth, and in this way become larger and more branched each succeeding year, until in old age they are very large.

6. **Ruminating Animals.**—Deer are ruminating animals, or, in other words, they chew the cud the same as oxen do. You may not have had an opportunity to observe this for yourselves in the deer, but no doubt you have watched the cow contentedly chewing her cud during her hours of leisure, and perhaps you have wondered at those lumps that rose in her throat soon after you saw her swallow her food.

7. **Digestion in Ruminating Animals.**—This process of chewing the cud is truly a curious one, and we must now try to find out what we can about the digestive apparatus

of the deer, the cow, the sheep, and other animals which possess a habit so peculiar.

8. These ruminating animals feed entirely upon vegetable substances, of which they take large quantities, and it is their custom to swallow their food hastily without much chewing, and then afterwards to raise it into their mouths and masticate it thoroughly at their leisure. They are enabled to do this by having a very complicated stomach, which is divided into four chambers, as shown in Fig. 247.

Fig. 248.—Reindeer Digging in Snow.

9. In the first place, we must notice that these animals have no incisors on the upper jaw, and the grass is drawn into the mouth with their long tongues and bitten off against the hard upper gum. When swallowed, the food

16

enters the first large stomach, or "paunch," where it is moistened with digestive juices, and then passes into the " reticulum," the inner surface of which is divided into small cells like a honey-comb. Here the food is pressed into little balls, which, by a process of muscular contraction exactly opposite to that of swallowing, return to the mouth to be eaten over again.

Fig. 249.—Travelling in Lapland.

10. The food, on being swallowed the second time, descends to the third cavity, and in doing so you will see that it must pass directly over the openings into the first and second stomachs; but the lip-like edges of these openings seem to have the power of selecting what shall be received and what shall be allowed to pass by.

11. This third cavity is called the "many-plies," from folds in the lining which resemble the leaves of a book. The fourth stomach, the "abomasum," supplies the gastric juice, and it is the true organ of digestion.

12. **Manner of Feeding suited to timid Animals.** — How admirably this manner of receiving food is suited to the

Fig. 250.—ANTELOPE.

shy and timid deer! They are so easily alarmed that it must be greatly to their advantage to be able to swallow their food rapidly and runaway to their shady retreats.

13. **Cloven-footed Animals.**—Those animals that chew the cud are also "cloven-footed;" that is to say, they have two toes encased in hoofs which have the appearance of one

Fig. 251.—THE KOODOO.

hoof that has been split into two equal parts. In addition to these toes, deer have two smaller toes at the back of the foot which seldom touch the ground, but still they are encased in dainty little hoofs.

14. **The Reindeer.**—Reindeer are confined to the extreme north of Europe and America, and they are the only species of deer that have been thoroughly domesticated. They are not only used for drawing sleds, but their milk and flesh supply the natives with food, and their skin is valuable for clothing. In Norway and Sweden large herds of reindeer are owned by the farmers,

who roam over the mountain districts with their herds to find summer pasture.

15. **What the Reindeer Eat.**—These strong, heavy animals eat scarcely anything but reindeer moss and lichens, which they obtain by digging with their fore-feet under the snow, and in doing so, as the hole grows deeper and larger, the animal is sometimes almost hidden from sight. They claim the privilege of searching their own food, and will not eat moss which has been gathered for them.

16. **Good Travellers.**—Reindeer are fine travellers, especially in cold weather. When the way is good, and not too hilly, they can travel a hundred miles in a day. Their

Fig. 252.—THE GAZELLE.

feet are well suited to walking upon snow, owing to the manner in which the hoofs separate in treading, and by the long, coarse hair growing between the hoofs. The foot may also be closed in such a way as to give a firm support in rocky places. This is the only kind of deer in which the females have antlers.

Fig. 253.—THE GRACEFUL CHAMOIS.

17. Proofs of the Extreme Heat and Cold in Past Ages.—
Remains of the reindeer have been found over the greater
part of Europe, and their presence in the same localities
that have at other times been frequented by the lion, the
rhinoceros, and the hippopotamus points to the fact that

these regions have been subject at different periods to the extremes of both heat and cold.

18. **Antelopes.**—Antelopes are especially abundant in Africa. These attractive animals are not classed with the true deer family, for they have hollow horns, and they do not shed them. The horns of some species of antelopes are dark and rich in color, and grow into beautiful shapes. The spirally twisted horns of the koodoo, for example, are very ornamental.

19. **The Gazelle.**—The best known of the antelopes is probably the gazelle, which is admired for the elegance of its form and movements, as well as for the mild expression of its "soft black eye." The gazelle roams through the wilds of Africa in large herds, and this gentle creature forms the ordiuary food of the lion and the panther.

20. **The Chamois.**—The chamois is a European antelope, living in flocks among the mountains, where it bounds with great ease over the rocky cliffs.

LIII.

CAMELS.

SUB-KINGDOM, VERTEBRATA: CLASS, MAMMALIA.

1. **Camels as Beasts of Burden.**—For centuries the camel has been used by merchants and travellers as a beast of burden to cross the sandy plains of Africa and Arabia, and long before the Cape of Good Hope was discovered

Fig. 254.—CAMEL.

treasures of gems and spices and the richly woven fabrics of the East were carried on the backs of these large, un- gainly animals to the shores of the Mediterranean Sea, that they might be distributed to all parts of Europe.

2. **The Structure adapted to Life on the Desert.** — Camels, from their peculiarities of structure, are well adapted to the life they lead on the desert. They are not only ruminating animals, but they can go for many days without water, being provided with a singular arrangement of

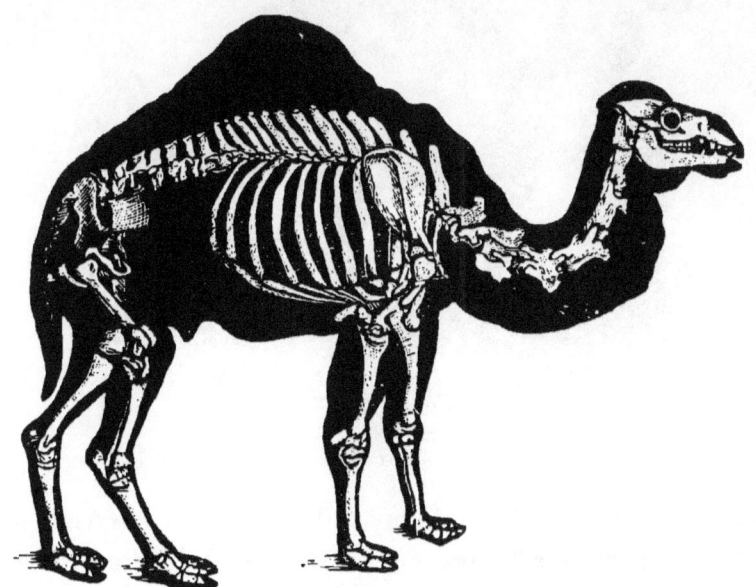

Fig. 255.—SKELETON OF A CAMEL.

cells in the first stomach, or paunch, which they fill when they have access to water, and keep as a reservoir for future use. Their two long toes rest on a broad, horny cushion which prevents them from sinking in the loose sand, and their nostrils can be closed at will to protect them from the fine particles of dust blown over the desert by the fierce winds that visit it.

3. **The Hump a Reserve Supply of Food.**—The large, unsightly hump on the back of the camel is not a part of the skeleton, as you will see by referring to Fig. 255, but it is only a mass of fat which slowly diminishes in size when

16*

the animal is on long journeys and food is scarce. The fat which is stored up in the hump is gradually absorbed into the blood, so the camel, in reality, carries a reserve

Fig. 256.—BACTRIAN CAMEL.

supply of food in this hump. The African or Arabian camel has but one hump, and is often called the dromedary. The Bactrian camel from Central Asia has two humps.

4. **Camels have been Domesticated.**—Camels belong exclusively to Asia and Africa, but they are no longer found in a wild state, man having appropriated them to his own uses. They are celebrated for their gentleness and patience, and are often required to travel with enormous burdens. When quite young they are trained to kneel down and receive these heavy loads, and in doing so they afford a most unusual picture of meekness and long-suffering.

5. **Their Uses to Man.**—The wealth of the Arab often consists in his camels, the uses of which are as various as those of the reindeer to the Laplander. The flesh and milk yield him food, the hair he weaves into clothing, the hides

he makes into sandals, saddles, and other useful articles. Seated with his family upon the back of his trusty camel, he is conveyed with long, shambling strides upon his weary journeys; his goods and chattels, piled up on the saddle and hung from its sides in indescribable confusion, are moved from place to place by the same conveyance.

6. **Camels raised in the United States.**—The experiment of raising camels in the United States has been tried with a good degree of success, and they now breed in Nevada.

7. **The Llama.**—The llama of South America is quite similar to the camel, though it is smaller and has no hump, and is in every way fitted for a mountain life. Each toe has a separate pad and a strongly curved hoof, which assists in climbing steep rocks, and the llama bounds over the cliffs of the Andes with the ease and activity of a goat. The teeth of camels and llamas differ from those of all other animals that

Fig. 257.—The Llama.

chew the cud, for these are the only ruminants that have incisors growing in the upper jaw.

LIV.

ELEPHANTS.

SUB-KINGDOM, VERTEBRATA: CLASS, MAMMALIA.

1. The Elephant's Trunk.—The great size of the elephant, and the remarkable trunk which it uses so nimbly, are sure to awaken our interest whenever we see these curious animals. Did you ever suspect that this wonderful trunk or proboscis is merely the nose of the elephant prodigiously lengthened out? for that is just what it is—a very long nose, which, oddly enough, serves also for an upper lip. The nostrils extend down through the whole length of the trunk, and above the openings into them there is a finger-like tip, which is used as a hand in picking up small objects.

2. The trunk is well supplied with muscles, which allow it to bend freely in every imaginable direction, and it is a most useful implement to the elephant, whose short neck and long tusks interfere with the usual manner of obtaining food and drink. Elephants cannot reach the ground with their heads to bite off their food, or to satisfy their thirst at the brooks and streams, consequently all their food is lifted to the mouth by the finger-like tip of the trunk, and, in drinking, water is sucked up into its hollow tubes. The end of the trunk is afterwards doubled up and placed in the mouth, and the supply of water it contains is emptied down the throat. Besides these impor-

tant uses, the trunk is also the organ of smell, of touch, and of defence.

3. **The Skull and Tusks.**—If you should have an opportunity to examine the skull of an elephant, you will find

Fig. 258.—African Elephants.

it to contain many hollow spaces which are filled with air, and which greatly reduce the weight of the large, clumsy-looking head. Its size and strength, however, are sufficient to support the huge tusks, which often weigh from one hundred and fifty to two hundred pounds. The tusks are the incisor teeth of the upper jaw, which continue to grow during the lifetime of the elephant, and sometimes reach a great length.

4. Their Home and Habits.—These large animals live in herds in the forests of tropical Asia and Africa. They feed upon grass and foliage, and seem to prefer the shade of the forests to the glaring sunlight, as they generally stroll out towards night.

5. The African Elephant.—The only species now living

Fig. 259.—INDIAN ELEPHANTS.

are the Indian elephant and the African elephant. The African elephant has great, flapping ears, and it is more fierce than that of India. It is hunted chiefly for its tusks, which yield fine ivory, and are therefore very valuable. The demand for tusks is so great that there is reason to fear these elephants will be entirely destroyed in order to supply the trade. The immense size of these living curiosities reminds us, in a way that no other animals do, of the huge monsters of various kinds that formerly dwelt upon our earth.

6. **The Indian Elephant.** — Indian elephants have mild dispositions, but if they are irritated they become furious and revengeful. It is stated that they can be easily tamed, no matter what their age or size may be, and in India they are used for many kinds of labor which require intelligence and skill. Strangest of all these employments is that of catching wild elephants.

7. **The Capture of Wild Elephants.** —To assist in capturing their fellows after having been deprived of their own liberty seems more than could be expected of these powerful animals; yet they enter into the labor with spirit, as if they understood the object to be accomplished and the best means to attain it. They urge on the reluctant ones among the wild animals which they are pursuing, pushing them forward if necessary, and if any are thrown down they kneel upon them and keep them upon the ground by their immense weight until they can be secured by ropes. It is claimed that these are not tricks taught to a few individuals, but that all working elephants in India are expected to possess such intelligence and sagacity.

8. **White Elephants.** — The "white elephants" about which we hear so much are merely albinos of the Indian species. In other words, they are of a lighter color than most of their kind. Generally they are far from white. Now and then an elephant is found with white spots on its skin, or sometimes the whole animal is of a light color; but whether it shall be considered a " white elephant " or not depends upon the decision of the people of Burmah and Siam. In these countries they are regarded as sacred,

and are claimed by the kings, who pay handsomely for them and keep them in royal style.

9. **Mammoths and Mastodons.**—Although but two species of elephants now remain, there is reason to believe that these giants were numerous in olden times. Mammoths and mastodons are no longer living, but their fossil remains in Europe, Asia, and America are found abundantly at the bottom of swamps, where the heavy beasts seem to have mired.

10. Several mammoths, covered with long woolly hair, have been found perfectly preserved in the frozen gravel of Siberia. Much of the ivory of commerce comes from Siberia, and it is obtained from these extinct mammoths.

LV.

LIONS AND TIGERS.

SUB-KINGDOM, VERTEBRATA: CLASS, MAMMALIA.

1. **The Lion.**—Among all the dreaded beasts of prey there are none so well calculated to inspire their victims with terror as the lion, on account of his great size, his

Fig. 260.—LIONS.

majestic bearing, his fierce countenance, and, above all, his terrible roar. We shall find by examining his formidable mouth that, like the mouths of other carnivorous animals, it is furnished with long, sharp teeth, well suited for tearing prey, and the tongue is roughened by horny points directed backward, which are of great assistance in scraping the flesh from the bones.

2. **The Characteristics of the Cat Family.**—Lions are often spoken of as belonging to the cat family, a group having strongly marked peculiarities. The members of this family are compact in form, without much fat; they are very strong, but, nevertheless, they are remarkably light upon their feet, and they tread upon the tips of their toes, the heel being raised from the ground, and the sole of the foot covered with hair like the rest of the body. These animals are nocturnal in their habits — that is, they prowl about at night—and they all spring suddenly upon their prey.

Fig. 261.—Foot of a Lion.

3. The toes are armed with hooked claws, which, when not in use, are drawn up within sheaths that they may not become blunted, and the same curious contrivance which draws in the claws provides also for darting them out again as soon as they are needed. You may not choose to examine the lion's foot very closely, especially as you can see this peculiarity with greater ease and more safety in the velvety paw of a cat. When the cat is in an amiable mood, its claws are so nicely folded in that the paw looks perfectly harmless; but if the cat should become vexed while you are watching it, and throw out its paw to scratch, the sharp claws will soon show themselves ready for service. The soft pads under the toes give these animals a stealthy,

noiseless tread, and they also serve to break the jar occasioned by their violent leaps.

4. **Lions confined to Asia and Africa.**—Lions are found only in the warm parts of Asia and Africa. They are so strong that they can carry large animals in their jaws, running and leaping the while as if they were not burdened with the heavy load. The male lion is ornamented with a bushy mane, which covers the shoulders, as well as the head and neck, and adds greatly to his majestic appearance. The lioness is much smaller than her mate; she has from two to four cubs at a time, and these are as playful as young kittens.

5. **Lions are Cowardly Animals.**—Although lions are noble-looking animals, they are by no means courageous in disposition. On the contrary, they are extremely cowardly, sleeping through the daytime and lying in wait for their prey at night. A dark, stormy night is their favorite time for starting out, consequently persons travelling through the countries frequented by lions seldom meet with them on their journeys. Those who have seen them, however, describe them as turning quietly round and trotting off when they find themselves discovered. They are much less feared by the natives than the ferocious tiger and the leopard.

6. **Redeeming Traits of Character.**—The tenderness which the lion shows to his mate, and his care to assist her in hunting food for their little ones, are redeeming traits in his character which are gladly recorded in his favor.

7. **The Tiger.**—The tiger, which is a native of Asia, is equal to the lion in size, and much more active in its movements. Its favorite manner of capturing prey is to conceal itself near some spot to which its victims are known to resort; then, with a terrific roar, it springs upon

Fig. 262.—TIGER ATTACKED BY A CROCODILE.

the unsuspecting animals and stuns them by its great weight.

8. There are few animals handsomer than the tiger. Its color is a reddish-yellow, striped with irregular bars of black, while the under part of the body is white.

9. **The Mimicry of the Lion and Tiger.**—In comparing the gay coat of the tiger with the uniformly dull one of the lion, we can but admire the mimicry of both. The coloring in these two animals is very different, yet the general appearance of each one answers the purpose of protection in its native haunts. Thus the dull, tawny fur of the lion is so much the color of the sandy desert that the king of beasts can hardly be distinguished at a distance, as he roams over its barren wastes. On the other hand, the

showy coat of the tiger serves equally well as a concealment in the jungles frequented by this stealthy animal,
since the stripes on the back have a general resemblance
to the tall, coarse grasses among which it hides.

10. **The Leopard.**—In the same way, when we consider
the habits of the leopard, we shall find that its conspicuous covering only adds to its security. Those handsome
spots which strike the eye so quickly in the caged animal
are not unlike the flickering shadows of the leaves in a

Fig. 263.—LEOPARD.

forest when the bright sunlight falls upon it; so that the
spotted leopard, when at home in its native woods, harmonizes well with the patches of sunshine and shadow by
which it is surrounded.

LVI.

SEALS AND WALRUSES.

SUB-KINGDOM, VERTEBRATA: CLASS, MAMMALIA.

1. Carnivorous Animals living in the Ocean. — Returning once more to the dwellers in the ocean, about which
there is always a peculiar charm, let us now study the

Fig. 264.—HERD OF SEALS.

seals and walruses. These are truly carnivorous animals, fitted for living in the ocean, and their small heads,
sloping shoulders, and plump bodies, gradually tapering

towards the tail, offer little resistance in gliding through the water.

2. The short limbs also are well suited to swimming; the hindermost ones, however, cannot move very freely, for they are set far back, and are so bound down by the

Fig. 265.—Harp-seal Mother and her Little One.

skin that they have the appearance of forming part of the tail. The close, thick fur and the layer of fat under the skin have their uses likewise, and protect these animals from the extreme cold to which they are exposed.

3. **Legends of Mermaids and Sea-nymphs.**—The prettily rounded head of the seal resembles the head of a dog, and when it emerges unexpectedly from the water its intelligent countenance and large, dark eyes are quite startling, and may easily have given rise to some of the legends of mermaids and sea-nymphs.

4. **A Sentinel on Guard.**—Seals feed chiefly upon fish, and

spend most of their time in the water, coming on shore to sleep in the sunshine and to suckle their young. One of their number is selected to act as sentinel while the others lie asleep on the rocks, and this sentinel keeps watch from some high point, ready to give warning if they are threatened with any danger.

5. **The Value of Seals to the Greenlanders.** — Seals are found in almost all seas, but they are especially abundant in cold climates. They form the main subsistence of the Greenlanders, who become very skilful in catching them. The flesh is used for food, the fat for fuel and for giving light, and the skin for clothing and for covering their huts and their boats.

6. **Seal-skin Fur.**—The beautiful seal-skin fur which we prize so highly as an article of clothing is obtained from the sea-bear of Alaska. Before the skin is dressed, the soft, velvety fur is hidden by long, coarse hairs which stick out beyond it; but when the skins are prepared for sale, these coarse hairs are pulled out, leaving only the short yellow fur underneath, which is then dyed to produce the favorite rich brown color.

7. **Sea-lions.**—Sea-lions are a large species of seal having external ears. They live in great companies in the Pacific Ocean, and towards spring they come to the shore to raise their little ones. The males land first, and take their positions upon the rocks to await the arrival of the females a little later. Fierce struggles then take place as they choose their mates from the herd.

8. Hundreds of these sea-lions may be seen clumsily dragging themselves over the rocks in San Francisco Bay. They have a loud, shrill bark, which they frequently utter, not only on the rocks but also in the water, when they rise to the surface for a fresh supply of air.

9. **Walruses.**—The walrus is much like the seal, except that it is larger and heavier, and it has two sharp-pointed

Fig. 266.—WALRUSES.

tusks formed of the upper canine teeth, which grow downward, sometimes to the length of two feet. These strong tusks must be of great assistance to the walrus in scram-

17

bling out of the water and mounting upon steep rocks and icebergs.

10. **Their Gambols.**—Walruses live in large herds in the Arctic seas, and are hunted for their blubber and their tusks. Assembled on the ice, they appear to have great sport rolling and tumbling heavily about, making the while a loud, bellowing noise. After their frolic the whole party generally fall asleep, except one walrus, which is left on guard. The polar-bear is their great enemy, and this expert diver and swimmer hunts them both in the water and upon land, so they are never secure from the danger of an attack.

LVII.

BEAVERS AND SQUIRRELS.

SUB-KINGDOM, VERTEBRATA : CLASS, MAMMALIA.

1. **Beavers.**—Not very long ago beavers were abundant in nearly all the wooded districts of North America, but they have become scarce, and are now found only in wild and unfrequented parts of the continent. Their hind-feet, as

Fig. 267.—BEAVER.

you see in the picture, are webbed for swimming, and they have a curious broad tail, flattened above and below like a paddle, and covered with thick skin.

2. **The Uses of the Tail.**—They have been said to use this tail as a trowel for plastering their dwellings, and

also for driving stakes, but authentic accounts inform us that the tail is used merely as a rudder in swimming, and as a support to the beaver while sitting up at its work.

3. **Beavers as Builders.**—An unusual degree of interest is felt in beavers on account of the skill which they display in building their homes and in felling timber for the construction of their dams. In the arduous labor of cutting down trees, the only implements used are the sharp, gnawing teeth; so we must examine the teeth particularly that we may see how they are enabled to perform such difficult tasks.

4. **The Teeth adapted to the Habits of Gnawing Animals.** —Beavers belong to the family of rodents, or gnawing animals; and as all these animals feed upon nuts, or the bark and woody stems of trees, they are supplied with sharp, chisel-shaped teeth, in order to nibble tough woody fibres. Indeed, the gnawing teeth form the strongest peculiarity of this order.

5. In each jaw there are two long, curved incisors, which are perpetually growing. The front surface of the teeth is covered with hard enamel, which does not wear away as rapidly as the body of the tooth behind it; therefore, the front part of the teeth always forms a sharp, cutting edge, as shown in Fig. 268. This wearing away from gnawing counteracts the continual growth, and keeps the teeth at about the same length. If by any accident one of the teeth is lost, its opposite neighbor has nothing to rub against and wear it off, and consequently it grows so long as to become a serious inconvenience.

6. A further provision for gnawing is shown in the arrangement of the mouth, the lower jaw being attached to the skull in such a manner as to slide backward and forward, thus aiding in the process.

7. **Beaver Dams.**—As has been stated, beavers show remarkable intelligence in building their homes, and they arrange them so that the entrance may be at all times beneath the water. When the home of the beaver is in a stream or lake deep enough to secure this important object, there is no necessity for a dam, or for the erection of houses, and their dwellings are then hollowed out in the banks. But if the stream is shallow, dams are needed to store up a sufficient quantity of water to conceal the entrance to their homes, as well as to prevent the possibility of its being blocked by ice.

8. In order to build these dams, trees must be cut down and dragged or floated to the spot; stones and lumps of earth are then brought to keep the timbers and boughs in

Fig. 268.—SKULL OF A BEAVER.

place, and everything is securely fastened. Twigs and pieces of wood are also stored up for winter food in case the beavers should be compelled to resort to such in-door fare.

9. **Their Sagacity in Cutting Timber.**—All the wood-cutting, as we have seen, is done with the sharp front teeth, and it is accomplished very rapidly. Sitting upon the tail and haunches, a single beaver gnaws a circle around

the trunk of a tree, going round again and again, gnawing the groove deeper each time. At length, when the trunk is cut nearly through, after examining it frequently, the careful worker nibbles only upon the side towards which it wishes the tree to fall, taking care to dash away at the first cracking of the timber, that it may not be crushed by the falling weight. The trunk is next cut into the desired lengths and dragged to the water, that it may be floated to the dam. When large trees are needed, the beavers are wise enough to select those that stand near the edge of the water, and they are careful to gnaw the trees in such a way that they shall fall into it, and thus save the labor of dragging them.

10. These logs are piled up to construct the dams, and the branches are plastered with mud and grass to form the house, which looks on the outside like a rough, irregular pile; still it is firm and well suited to the needs of the beaver.

11. **A Beaver Family.**—A beaver family rarely consists of more than twelve inmates. Frequently the families scatter in the spring and live separately during the summer; but before cold weather comes they gather together again, and every one, both large and small, helps in repairing the dam and the dwellings, which have suffered from neglect during their absence.

12. **Squirrels.**—The graceful squirrels, jumping from branch to branch among the trees, with their long, bushy tails curled up over their backs, and their large ears erected to catch the faintest sound, are attractive little animals belonging to the same family as the beaver. The handsome tail of the squirrel is more than a mere ornament, for it aids in leaping, and also makes a warm wrap at night.

Fig. 269.—SQUIRRELS.

13. **The Storehouse of the Squirrels.**—Squirrels are careful in summer to lay up a supply of nuts for winter use, often carrying as many as four or five acorns at a time in the curious cheek-pouches with which some species are

Fig. 270.—FLYING-SQUIRRELS.

provided. A hollow tree is generally selected for their storehouse, and the squirrels pass the cold weather in the same sheltered domicile, but their nests are commonly built in the tree-tops.

17*

14. A Squirrel Nibbling its Nut.—When eating its favorite nuts, the squirrel sits upright, holding the nut daintily in its fore-paws, and turning it from side to side while it gnaws away at the shell with its sharp little teeth.

15. The Flying-squirrel.—The flying-squirrel is one of the prettiest of the squirrels. It has really no power of flying, but there is a furry skin extending from the fore-legs to the hind-legs, which, together with the broad tail, acts as a parachute, and supports the active creature for a time when it leaps into the air. The flying-squirrel is seldom seen, even by those in whose neighborhood it lives, because this shy little nut-gatherer ventures out mostly at night.

LVIII.

BATS.

SUB-KINGDOM, VERTEBRATA: CLASS, MAMMALIA.

1. Bats are True Mammals.—Bats are the only mammals that can truly fly, and in studying them we must not lose sight of the fact that, although these animals lead a life similar to that of birds, yet they are, in reality, mammals. They are covered with soft fur instead of feathers; they have large ears and noses, with a distinct pair of nostrils;

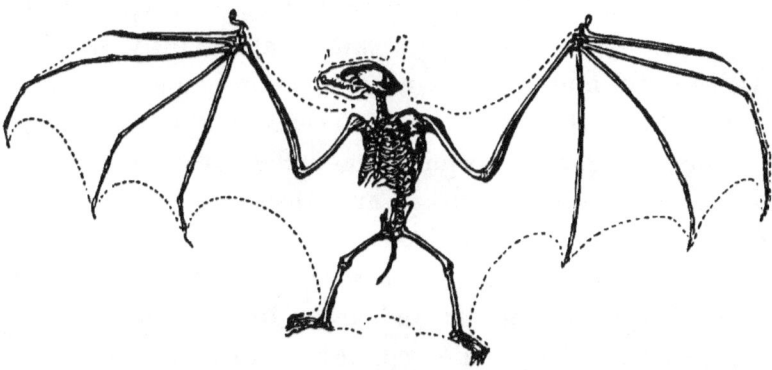

Fig. 271.—SKELETON OF A BAT.

their mouths are furnished with two rows of sharp-pointed teeth, and their eyes are protected by eyelids and eyebrows. Apparently these bright black eyes distinguish objects very imperfectly in the broad daylight, and they seem to be better fitted for the dim twilight or the darkness of night.

2. **The Wings of the Bat.**—We may see in the skeleton of the bat (Fig. 271) how very much the arms and fingers are lengthened out to form what we usually call the wing. These long bones support the delicate skin which is spread over them in somewhat the same way that an umbrella frame supports the silk which is stretched over it. This silky skin connects the fore limbs with the hind limbs, and generally extends to the tail, making an excellent substitute for wings. As an additional preparation for flying, there is a keel on the middle of the breastbone for the attachment of flying muscles, an arrangement similar to that which we noticed in birds.

3. **Their Delicate Sense of Touch.**—Bats fly rapidly, although their movements are awkward and aimless, consisting of innumerable darts and sudden turns. Their sense of touch, especially upon the smooth skin of the wings, is very acute. This skin is abundantly supplied with nerves, and, by its sensitiveness, it helps the bat to discover the presence of small insects in the air. Large numbers of gnats, mosquitoes, and flies are devoured by the bats as they fly hither and thither in the twilight hunting for this kind of food.

4. **How the Bats Live.**—These singular animals live almost entirely in the air, and are quite helpless when on the ground. They build no nests, and sleep with their heads hanging downward, suspended by means of hooks on their hind-legs. In this position they pass their days in dark caves and crevices, or, not finding these, they take refuge under eaves or around old church-steeples. They spend the winter in a torpid state, clinging securely to some such places as have been described.

5. **The Young Bats.**—Bats generally have two little ones at a birth, and these young bats cling so closely to their

mother's breast that she can fly with a pair of them attached in this way and not appear to feel their weight. The skin which covers her tail is folded up over them, and her young family is thus wrapped in a safe pouch while she flits about in search of food.

6. **The Vampire Bat.**—The vampire bat of South America is a large bat which measures two and a half feet in width when its wings are spread. It feeds upon insects,

Fig. 272.—Bat.

and has the reputation of killing larger animals, and even human beings, by piercing a small round hole and sucking the blood while its victims are asleep. Such cases rarely occur. The bat does not draw enough blood to cause death, but it has been suggested that the animals thus attacked may be weakened beyond recovery by the blood continuing to flow silently from the wound after the bat has satisfied its appetite and gone away.

LIX.

MONKEYS.

SUB-KINGDOM, VERTEBRATA: CLASS, MAMMALIA.

1. **Four-handed Animals.**—Monkeys are often spoken of as four-handed animals, because their feet as well as their hands are fitted for grasping objects. The formation of the foot is peculiar in having the great toe separated from the other toes, so that it can be brought opposite to them in much the same way as our thumb folds upon the fingers, and in consequence of this arrangement the feet may be used as a second pair of hands.

2. **The Home of Monkeys.**—Monkeys are particularly numerous in the great tropical forests. They feed upon fruits, young birds, and birds' eggs, all of which they find in the sheltered tree-tops; therefore they have little occasion to come to the ground, and they pass most of their lives among the leafy branches, running, jumping, and swinging from tree to tree.

3. **The New World Monkeys.** — The monkeys of this continent are confined to Central and South America, and they are known by the general name of New World monkeys. They differ in many respects from the Old World monkeys. They are generally small animals, the nostrils are far apart, and are placed near the end of the snout. Most of them have long, prehensile tails, and are great climbers.

4. **The Difference between the New World and Old World Monkeys.**—There are a few strongly marked characteristics by which these two classes of monkeys may be readily distinguished. Thus, you may feel quite certain that any monkey with a long tail which it can curl up at the end for the purpose of taking hold of things belongs to

Fig. 273.—White-throated Sapajou.

one of the American species, whereas, on the other hand, one that has bare seat-pads may be recognized as having come from the Old World. It is generally the American monkeys that are seen dressed in little jackets and begging pennies for wandering musicians.

5. Some of these New World monkeys are very intelligent, and so droll and full of mischief that their pranks

are quite amusing. Let us choose from among these the spider monkeys, with their slender bodies and very long tails and limbs. This long tail is so strong that it answers the purpose of a fifth hand, and is a valuable assistance

Fig. 274.—MANDRILL.

in climbing and jumping from one tree to another. The active little monkey often twists the end of its tail around a branch, and with no other support swings freely in the air.

6. The Old World monkeys are much more highly developed than their playful relations on this side of the ocean. Their nostrils are placed nearer together, and open downward more like our own. The arrangement of the teeth is similar to that of man, although the front ones are large and prominent, and they are uneven in length. Many of these monkeys are entirely without tails, and in those species that possess a tail this latter appendage is never prehensile.

7. The Gibraltar Monkey.—The only wild monkey living in Europe at the present time is the celebrated Gib-

raltar monkey, which roams about the Rock of Gibraltar, and is carefully protected by the government.

8. **Baboons.**—Baboons are found in all parts of Africa, and they are the fiercest and most dangerous of the monkey tribe. Their long snouts give to the head somewhat the shape of a dog's head. They have cheek-pouches, in which they stow away their food, and the hard pads of bare skin on which they sit are usually of some bright color. Mandrills are a large variety of baboons, with swollen cheeks ornamented with red and blue stripes, and their appearance is rendered exceedingly disgusting by other patches of gaudy color. .They associate in bands,

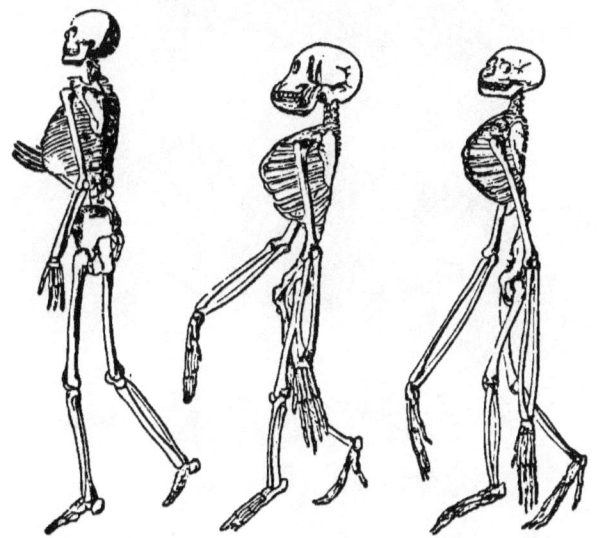

Fig. 275.—Skeletons of Man, Chimpanzee, and Orang.

and are so strong that when assisted and encouraged by their fellows they do not hesitate to attack the elephant.

9. **The Group of Monkeys known as Apes.** — Gibbons, orang-outangs, chimpanzees, and gorillas are called apes because their structure approaches more nearly to that of

man than is the case with any other animal, and on this account they have a peculiar interest to the student of natural history.

10. **Gibbons.**—Gibbons live in troops in the forests of India and the adjacent islands, and are mostly led by one male, who is their chief. They are awkward-looking animals, with long arms that extend to the ground when they stand upright.

Fig. 276.—FEMALE ORANG-OUTANG.

11. **Orang-outangs.**—The ugly orang-outangs live on the islands of Borneo and Sumatra. Here they inhabit the densest forests, and are commonly known as "men of the woods." When fully grown they reach the height of four or five feet; and although they are heavy, clumsy creatures, yet they spring about among the branches with great ease and rapidity, seldom coming down from their haunts

unless it be to obtain drink or to shuffle off to some new locality.

12. They do not usually walk erect, but in moving along the branches of the forest they often choose an upright position, and support themselves by taking hold of the boughs overhead. Orangs build a broad nest low down in the trees by piling leafy branches loosely upon each other without interweaving them. Here they sleep at night, and do not leave their nests until the morning sun has dried the dampness from the surrounding leaves.

Fig. 277.—CHIMPANZEE.

13. **Chimpanzees.**—Chimpanzees are natives of Western Africa. They have no hair on the hands and face, and none on their large, rounded ears. Altogether their general resemblance to man is decidedly grotesque. Their arms

Fig. 278.—GORILLAS.

are shorter than the orang's, still they fall below the knee.
They can walk erect, although they seldom do so, their
habit being to bend forward and rest upon their hands as
they move about.   They live in companies in the woods,

and form nests among the branches near the ground. When tamed, chimpanzees have sometimes been taught to eat their food with a spoon at the table, and to imitate some other customs of human beings.

14. **Gorillas.**—Much larger and more ferocious than the chimpanzees are the gorillas of Western Africa, which are often found six feet in height. These strong animals live in bands, and build nests which are occupied only at night. Gorillas are now generally considered to be the most highly developed of the apes.

LX.

MAN.

SUB-KINGDOM, VERTEBRATA : CLASS, MAMMALIA.

WE have now traced the gradual development of animal life upon our earth, from the simple forms to the extremely complex ones, and our only remaining subject is man, the acknowledged head of the animal kingdom, the study of whose physical and intellectual nature forms separate branches of science.

The habitual position of man is erect; the lower limbs are used only for walking and for supporting the weight of the body. The arms are much shorter than the lower limbs, and they terminate in a hand which is admirably adapted to ministering to his needs, and to performing all the delicate operations which beautify and enrich his life. Above all, man is gifted with the power of speech, and with mental and moral faculties capable of the highest cultivation.

INDEX.

THE END.